SpringerBriefs in Applied Sciences and Technology

For further volumes:
http://www.springer.com/series/8884

Saurabh Kwatra · Yuri Salamatov

Trimming, Miniaturization and Ideality via Convolution Technique of TRIZ

A Guide to Lean and High-Level Inventive Design

 Springer

Saurabh Kwatra
Institute of Innovative Design
Krasnoyarsk
Russia

Yuri Salamatov
Institute of Innovative Design
Krasnoyarsk
Russia

ISSN 2191-530X   ISSN 2191-5318 (electronic)
ISBN 978-81-322-0736-8 ISBN 978-81-322-0737-5 (eBook)
DOI 10.1007/978-81-322-0737-5
Springer India Heidelberg New York Dordrecht London

Library of Congress Control Number: 2012943365

Printed on acid-free paper

Springer is part of Springer Science+Business Media (www.springer.com)

Dedicated to *Gudiya*

Foreword

TRIZ as a word and an innovative methodology is still an enigma with the engineering and design community in India. Creative processes are generally believed to be intuitive not systematic in nature. TRIZ bellies this belief and therefore is not easily accepted. Another factor for its not being well known is because of its Russian origins, as Russian technical literature was not easily accessible and the availability of technical literature from the West, particularly from the USA and the UK was overwhelming. A small minority from the technical community of India, however, is aware of TRIZ, but the practice of this science, if we may call it, is not very prevalent, main reason being that there are hardly any practitioners and trainers available.

Long time ago, when I attended a lecture on TRIZ for the first time, I thought it was a very bureaucratic and tiresome way of finding solutions. But later, after getting a little more educated, I realized it takes the uncertainty out of the problem solving process and lends itself to the corporate way and more participative way of working. TRIZ has taken the mystery out of the creative process. Engineers, therefore, can feel more confident of using this methodology and more certain of getting results.

With the publication of this book, a very real need for creating awareness about TRIZ will be fulfilled. It can help trainers to develop courses in TRIZ training and help to develop more practitioners, which can have a cascading effect in a positive way. The latest phenomenon/trend of setting up innovation groups/departments within Indian corporates is likely to give boost to the use of TRIZ as innovative methodology. This book will be handy guide for the managers of these departments. It will also be a good reference material for the students of creative process and innovation.

The book is thankfully full of highly communicative illustrations, which makes the book easy to understand. The examples very well explain the laws of TRIZ, and are highly inspiring, giving the reader, a feeling, "How I could not think of this before", making this book very credible. However, I would have liked to have some examples from the local context, but I am sure, we will have these in the next edition, as this first edition will encourage more people to adopt this method of problem solving.

I must congratulate the authors Saurabh Kwatra and Yuri Salamatov and also the publishers, Springer India for bringing out this book, which I am sure will fulfill the long felt need not only in India but in other parts of the World as well, where innovation and structured invention is valued.

Industrial Design Centre, IIT Bombay   Prof. K. Munshi
Mumbai, India

Preface

In design projects at all levels, bottlenecks are reached often. Conventional resources in hand to move ahead are brainstorming, trial and error, consultation with a senior; all these techniques often lead to a compromise solution. In TRIZ terms, these are inventions, but of lower levels. In contrast, TRIZ is ever itching to notice contradictions everywhere: administrative, technical, and physical. First type, viz. administrative appear in TRIZ application to social, political, or economic systems. It is technical contradiction, called TC and physical contradiction, called PC that is of particular interest to TRIZ for engineers, technologists, and scientists. They are worth exemplifying. If wings of plane are broadened, lift increases but drag increases too. If wings of plane are narrowed, lift decreases but drag decreases too. In this case, 'width of wing' is a physical characteristic (one of important dimensions) of airplane, while lift and drag are system properties (actually forces in aerodynamics). If we frame this challenge without 'width of wing', we get this: if lift improves i.e. increases, drag worsens, i.e. increases; if lift degrades i.e. decreases, drag improves i.e. decreases. Little need to mention, that lift is a desired property while drag is an undesired one—reduction of latter is effectually betterment. We call this TC. In TC, improvement of one system property inevitably leads to worsening of another system property. The same challenge could have been stated in terms of wing span alone: wings of plane must possess large area and small area together. We call this PC, wherein one physical characteristic like mass, size, length, temperature must have 'dual' values simultaneously. Readers must be curious to know how TRIZ proceeded in this aeroplane case. Shorting the process of TRIZ application, TRIZ works like this: instead of shying away from contradictions, TRIZ strengthens them. Then it applies its tools like Altshuller's matrix followed by inventive principles, principles of PC resolution like separation in time or space, laws of technical systems' evolution, ARIZ, etc. The laws of technical systems' evolution are one of strongest and universal techniques of TRIZ. Aim of entire TRIZ treatment is to reach a witty so-called convoluted technical solution. The technical system is leaned, stripped of useless weight yet made more capable of performing function it is supposed to. This simultaneous enhancement of useful effects with shrinking or envisioned vanishing of harmful effects, called convolution is so admirable that we call it idealization. This book studies this phenomenon, its natural occurrence in evolutionary graph of systems and most importantly methods to pre-pone it and to apply it to particular systems of interest with special focus on set of system variables. These methods fall under 'trimming'—a highly commercial term finding prime place in business plans of innovative design of factory processes, product manufacturers, and industrial corporations. Before closing, solution to airplane lift dilemma: foldable wings were invented by man: separation on condition has occurred. During take-off and landing when lift is supreme and necessary under low speeds, flaps are opened. During cruise when high speeds can easily provide lift and drag becomes a strong evil, they are closed.

It would be injustice on our part, if thanks are not given to Aninda Bose of Springer for untiring sincerity, Oleg Kraev of Institute of Innovative Design for initiating idea of a book in my mind and last but not least to Prof. K. Munshi of IDC, IIT Bombay for writing such a catchy foreword.

Saurabh Kwatra
Yuri Salamatov

Contents

1 Laws of Technical Systems' Evolution 1
Existence and Position of Laws in Engineering................. 1
Concise Treatment of Laws 1st Till 6th...................... 2
MUF, MDE, Ideality, and Ideal Final Result 2
1st Law–Law of System Completeness 4
2nd Law–Law of Energy Conductivity in Systems 9
3rd Law–Law of Coordinating System Rhythms
or Law of Harmonization................................. 12
Not a Law but a Strongly Observed Trend: Dynamization 14
4th Law–Law of Increasing the Degree of Substance-Field
Interactions of Technical Systems 18
5th Law of Transition from Macro- to Micro-Level.............. 20

2 Origins of Convolution................................. 23

3 Idealization and Convolution: Two Sides of Same Coin? 31
Emergence of Convolution in TS Evolution: Concise Treatment..... 31
Emergence of Convolution in TS Evolution: Extended Treatment.... 34
Tracing Ideality in Expansion-Convolution Waveform............ 43
   Case Study 1: TS is Refractrometer..................... 45
   Case Study 2: TS is Mirror for Powerful Laser............. 49
Convolution: A Graphical Perspective 56

**4 Four Types of Convolution: Miniaturization
Embedded in 2nd Type** 61

5 General Scheme of TS Evolution in History of Technology 75

6 Convolution and Trimming via Convolution 79
Case Study 1: Direct Invention of a Highly Convoluted TS 80
Case Study 2: Trimming of an Existing TS 82
Trimming of Street Light Pole 87
Trimming of Lighted Screwdriver: Invention of Nano-LED
Guided Screw-Driver..................................... 88
　　Development of SS(l) 91
　　Modification of Parts of TS to Accommodate SS(l) 91
Miniaturization of Existing Portable Gauss Meter 92
Lighted Kite Flying: Convolution 94
Trimming of Washing Machine Using Innovative Design Methodology   95

Appendix: Teaching Convolution in Classrooms 103

Chapter 1
Laws of Technical Systems' Evolution

Existence and Position of Laws in Engineering

TRIZ is abbreviation of *Teoriya Resheniya Izobretatelskikh Zadatch* in Russian. In English, it translates to 'Theory of Inventive Problem Solving'. TRIZ was founded by great Soviet engineer, Gerald Altshuller, beginning 1946. For some, TRIZ is a powerful design methodology, others use it as creative imagination booster and few others use it as tool to overcome deadlock situations faced in technical progress. A popular nickname for TRIZ in west is 'patent on demand'. After searching, sorting, classifying, and profoundly studying millions of patents, Altshuller came to a startling conclusion: technical systems evolve while adhering to some laws. If these laws are generically defined and applied to a particular technical system (TS)[1,2] of interest to innovator purposefully, then similar technical systems refine quicker. These 'laws of technical systems evolution' (LTSE) constitute a very significant portion of TRIZ content. The laws are:

1. Law of System Completeness
2. Law of Energy Conductivity in systems
3. Law of Harmonization
4. Law of Transition from Macro- to the Micro-Level
5. Law of Increasing the Degree of Substance-Field Interactions
6. Law of Transition to the Super-system
7. Law of Irregularity of System's Parts Evolution
8. Law of Increasing Ideality of Technical Systems

---

[1] Technical System or TS will soon be defined unambiguously. Till then any mechanism, machine, process, etc. may be termed TS. TS can be used in singular or plural sense–Technical System or Technical Systems will henceforth be TS. 'System' is short form for TS.
[2] For simplicity, TS can be used in singular and plural senses.

S. Kwatra and Y. Salamatov, *Trimming, Miniaturization and Ideality via Convolution Technique of TRIZ*, SpringerBriefs in Applied Sciences and Technology, DOI: 10.1007/978-81-322-0737-5_1, © The Author(s) 2013

Concise Treatment of Laws 1st Till 6th

he laws of TS evolution reveal considerable, steady, and repeated relation between elements inside of system and elements of exterior environment in the process of progressive development, i.e. of transition of system from its one state to another with purpose of increasing of its useful function. It is an undisputable and universal fact that any TS builds up its package of useful function(s) qualitatively and/or quantitatively over time. After all, sole aim of TS coming into existence is promise of its delivery of useful function(s) sought by man.

Laws were revealed during analysis of large groups of facts like inventions from patent fund, historic-technical researches. However, in engineering the laws act as spontaneous force and one can never be confident that steady, considerable, and serious (as opposed to casual) system relations are at play in the selected group of facts picked over a short period of time. Any reasonable period of time chosen to discover laws will be short when placed alongside entire inventive age of human civilization. That is why cognition of laws goes by the method of successive approximation: deliberately choosing stronger inventions (technical solutions), revealing of principles of technical contradictions during resolution, selection of the steady combinations of principles and physical effects, and finally standard steps in the development of technical systems. All investigation phases are prone to subjective factors like individual approach, estimation, absence of quantitative criterions, etc. Hence suppression or cancelation of such errors is also mandatory.

MUF, MDE, Ideality, and Ideal Final Result

Main useful function (MUF). Every TS is characterized by its MUF. A TS with nil MUF simply would not exist because man would not create it in first place. Primarily, man is interested in fulfilment of a particular function. When he finds his limited abilities cannot deliver that function, he looks at technosphere[3] with hope. He is then compelled to assemble a new TS that performs needed function. The needed function of human translates to MUF of TS. In case a TS performs multiple functions, MUF can be substituted by $Fn\sum(MUF)$. For simplicity of expression, MUF is often used universally, both in singular as well as plural sense. Any TS is always ready for its MUF increase.

Mass, dimensions and energy consumed (MDE). Every real TS is characterized by MDE. Another close term is weight, size and energy consumed (WSE); both MDE and WSE are interchangeably used. As obvious it is, MDE is an undesired characteristic of TS; TS is always (or should be) prepared for its reduction.

---

[3] The entire human-built world is called technosphere; term in contrast to nature created biosphere.

The most comprehensive, profound criterion for measuring progressive changes in development of any technical system is factor of *Ideality* or just Ideality. Ideality of a TS, I(S) is defined as below:

$$I(S) = \frac{F_n \sum MUF}{MDE}$$

When greater precision is required, MDE is substituted by Harmful Effect(s) or HE. Once again, HE is used in dual sense, singular, and plural.

HE = MDE + cost of production of TS + pollution caused by TS + other losses.

The 8th law, viz. law of TS idealization presents increase of Ideality as ultimate *aim* of all TS. It is the most important of all laws. All remaining seven laws are in fact concrete realizations of this 8th law on different stages of TS developments. Readers are cautioned not to over-read meaning of this, viz. 8th law. This law does not assert continuous rise of Ideality with TS evolution. It only establishes increase of I(S) in TS evolution in the long run. I(S) of a TS sequences as: a little decrease or constancy followed by substantial increase followed by a little decrease or constancy followed by substantial increase. Net rise is assured.

What is the upper limit of I(S) set in minds of TS or mind of its innovator? Infinity! What would be the physical state and configuration of TS when I(S) $\to \infty$ or is high enough? Answer: Ideal Final Result (IFR) state, called IFR in short.

TS strives to achieve its IFR state. Of course, it is easier said than attained. As TS attempts progressing toward IFR, it faces technical problems, mostly of inventive kind. The inventive problem is in turn is narrowed down to a technical contradiction (TC). Removal or relaxation of TC supplies required technical solution. Many a times, we are stuck with a TC; still that does not prevent us from formulating and appreciating terminal goodness of TS, viz. IFR.

There exists a scheme to formulate IFR easily and universally. One of elements of an 'ill' place of system or an environment itself eliminates harmful (unnecessary, superfluous) action, keeping ability to make useful action.

Here the magic word is 'itself', i.e. without participation of the person, without inflow of additional energy, without clubbing new subsystems, without interference of super-system. 'Itself' is used in absolute sense—without anything. In reality, it is impossible to achieve such a result. IFR in most cases is just a leading light, allowing an inventor to be drifted to best plausible solution? The aspiration to come nearer to IFR cuts all solutions of lower levels. It snaps them at once without an enumeration of possibilities. IFR and a small set of variants close to it remain. Inventive solutions of higher levels 'pay' a lower 'price' for changes in system while achieving more desired effects simultaneously.

An example from pharmacy industry demonstrates how a high level inventive solution close to IFR is wittingly cheap (in terms of costs paid). A machine manufacturing round 'tablets' pressed from a powder is showed in Fig 1.1 below. Powder is filled in small bunker. About 20–25 tablets are produced from each

Fig. 1.1 Tablet separator

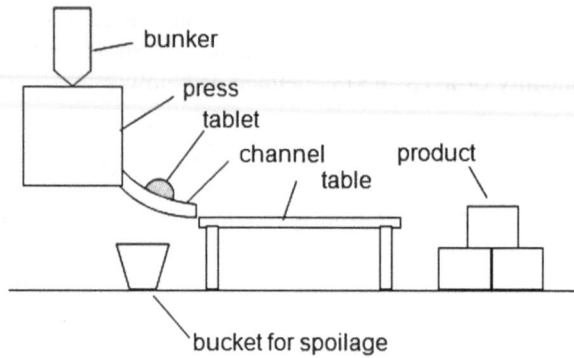

batch of powder. Every tablet, in standing edge position, is rolled down a channel on the table for packing into boxes.

But as a rule, last tablet is made incomplete; there is no powder to finish it. It cracks, crumbles, litters the table, messing the good lot. Workers then throw this 'trash' in a dustbin. One way to avoid littering is it to let alert worker 'catch' this broken tablet before it is received on table. We now formulate as: let the defective tablet be 'itself' detected and trashed. The corresponding inventive solution generated involves a slight shifting of either table or channel away from each another. A small gap does the job. The bucket is now repositioned vertically below this important gap. The spoilage itself slips into the bucket while the whole tablets roll with a certain velocity to skip gap and land onto table.

1st Law–Law of System Completeness

TS must be complete itself to be able to perform MUF. It ought to contain four parts for sure: working unit, transmission, engine, and control unit. Figure 1.2. These parts can be identified by their key roles given below:

Product: that which is processed. Processing can be displacement, transformation, production, modification, improvement, detection, safeguarding, measuring, etc. Product is considered outside TS, though inclusion in TS is not a grave error.

Working Unit: place where the energy inputs
Transmission: through which energy flows
Engine: source of energy
Energy/Power source: source of energy for engine
Control Unit: place that signals operation of all parts of system

Few salient comments: working unit is often akin to tool of a machine or system; engine receives energy from energy source in probably disorganized, undirected, or unusable form and converts this energy into more organized, directed, or usable form. Engine then supplies this 'better' energy.

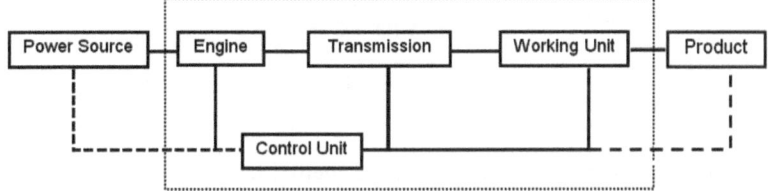

Fig. 1.2 Schematic diagram of TS

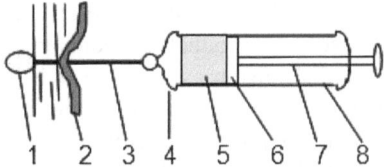

Fig. 9. Syringe - technical system.

1 - liquid,   2- skin,   3 - needle,   4 - cannula, - liquid,   6 - piston,   7 - rod, 8 - cylinder.

Fig. 1.3 Medicinal syringe as TS

Example 1: TS is Rifle. MUF is hitting a target.

Preliminary investigation is often erroneous as below.

What is processed? Bullet. Bullet becomes product.

What is the place where the energy inputs? Bullet. Bullet becomes working unit.

Through what energy flows? Barrel. Barrel becomes transmission.

What is the source of energy? Powder gas. Powder gas becomes engine.

What is the source of energy for engine? Chemical reaction, specifically explosion of gunpowder, more exactly explosive energy released becomes energy source.

Above analysis contains two mistakes:

1. Product is defined not correctly. Product is target, outside of technical system. Bullet is working unit.
2. Transmission is also defined incorrectly. Energy is transmitted by powder gas to working part, viz. bullet. Powder gas becomes transmission. Of course powder gas is simultaneously also the engine: powder gas transforms the explosion energy into forward movement. Barrel is part of engine too; it directs stream of powder gas.

Example 2: TS is Syringe. MUF is not mentioned (Fig. 1.3).

As usual, we begin with first question, 'What is processed?' Multiple answers are outcome.

Fig. 1.4 Judging WU in MUF

(a) Body (organism) is processed by liquid medicine
(b) Liquid is pushed out, i.e. processed by piston
(c) Needle pierces, i.e. processes skin

Then how to choose product? Best way to proceed is to keep human being or living organism out of ambit. The middle choice is left out now: Liquid is processed by piston. Liquid becomes product. It seems piston will become working unit. Let us flawlessly establish piston as working unit.

MUF of TS is intimately related to product. Liquid has been established as product above. Displacement of product, specifically its entry under the skin, is obvious MUF. Working Unit, hereafter called WU, is characterized by two strong features. It is the part to which energy from engine is delivered. It is also the part which fulfills MUF of TS. Figure 1.4 illustrates this.

What part in syringe fulfills MUF, i.e. displaces liquid? It is a piston. What is the place where the energy inputs? Piston; from there to liquid. Therefore, WU is piston.

The remaining parts: Transmission is piston-rod, Engine is piston-rod as it aligns the motion of finger transforming not-always organized motion of finger into unidirectional movement of piston-rod. Energy source is the hand of man. Cylinder is also a part of engine; it directs the motion of piston-rod and piston.

What about needle? It is WU, but of another auxiliary system—subsystem for piercing of skin. In this subsystem, skin becomes product being processed (pierced), needle becomes WU, cannula and cylinder are transmission, and hand of man is engine as well as energy source. MUF of this subsystem is piercing of skin. Any doubts in above demarcation are shed-off witnessing a nurse injecting a patient. She holds upper part of cylinder and delicately pushes the needle into skin. Force to needle comes via this route—upper part of cylinder to lower part of cylinder to cannula to needle. Since hand is engine and energy source both, we must mentally sub-divide hand if feasible. Upper part of hand acts more as energy source whereas lower part of hand (fingers) acts as engine; former supplies raw muscular power to latter. Latter converts it to a carefully directed and delicate force. Latter then delivers this refined, controllable force. This explains why a strong and heavy nurse can give soft painless injections.

Auxiliary systems are first to undergo alteration or even disappearance in process of TS development. Modern injection pistols have no needle. They are often used in mass vaccinations to control epidemics.

Problem 1: Underground pipeline built 50 years ago was recently investigated. It was found that it had still not lost durability. Only minor cracks and other small defects on several sites were discovered. Pipeline was in sections, each section about 100 m long with a well on either side. It was decided not to replace these pipes, and to cover them from within by polymer material. Idea sounded simple: to cover polymer sleeve outside by glue, to move it inside damaged pipeline between two consecutive wells, and then to fill it by water or air under pressure and wait until glue fastened. But first attempts to move sleeve turned out to be unfortunate. In the first instance, a cord was used to push sleeve from one end to another. This procedure was effective only in the beginning; soon sleeve got crumpled, wringed, stick in pipe and glue began to be grasped. Time was wasted in dragging out parts of sticky sleeve. How was polymer sleeve pushed through? How was it ensured that it moved along the pipe axis? How was it ascertained that it uniformly and accurately stuck to the inner surface of pipe?

Hints to solution:

1. There is no system. We should synthesize TS. MUF of TS needs to be stated. Parts of TS like WU, transmission, engine, etc. need to be pointed. Product need to be stated.
2. Few logical conclusions from problem conditions: For sleeve to stick to internal surface of pipe, it should contain glue. But for sleeve motion to be unhindered in pipe, sleeve should not contain glue. Contradiction emerges. We need to resolve it? From where should glue appear at required moment?
3. How is sleeve translated in pipe? It is next to impossible to draw sleeve from contrary end of pipeline. Final Technical Solution in Fig. 1.5.

Problem 2: *This is problem #47 of book titled 'The Right Solution at the Right Time' by Dr Salamatov.* 'Exhaust pipes of trucks have a large diameter and should be capped before parking. Otherwise the pipes get clogged with dirt or solid objects. Removable caps often get lost. Flapping caps are ineffective, because their fixing hinges get covered up with dirt and soot and cease to work properly. Can you think of a more reliable cap?'

Previous attempts to use mechanical field of exhaust gasses, as in flapping caps, had failed. We need to think afresh. Only one substance, viz. cap exists. It is the product to be treated, i.e. opened and closed. Entire TS needs to be assembled. There is no engine, no transmission, and no control unit. The control unit should issue a command to open cap when exhausts come out and to close it when there is no exhaust. Accordingly, the system should be controlled by gas i.e. by its emergence and disappearance. We now attempt using heat energy of exhaust gasses as energy source. There needs to be an engine that can transform input heat energy from energy source into mechanical energy as output. This mechanical energy would then open and close cap. The easiest way is to use a bimetal plate. Heat will bend plate which in turn would open cap, while cooling the cap will shut it reversely. In this case, TS consists of:

(a) thermal field of hot exhaust gasses as source of energy and control unit
(b) bimetal plate as engine, transmission, working unit, and a
(c) cap as product, of course outside TS

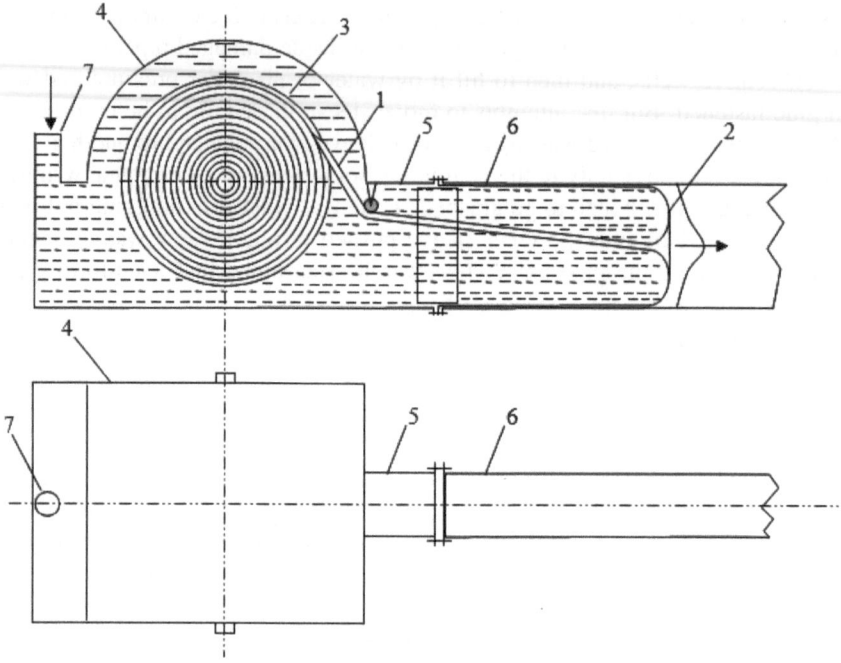

Sleeve made of film 1, having the glue on the internal side 2, is twisted on the drum 3. Drum 3 is placed in the waterproof reservoir 4. The end of sleeve 1 is fastened along the perimeter of socket 5 and tube 6. Then water is delivered through the choke 7 into reservoir 4. Water turns out sleeve and press film to internal surface of the tube.

Fig. 1.5 Soviet patent as technical solution

System can be further developed introducing nitinol, a nickel, and titanium alloy with memory shape effect. The cap gets twisted upon heating and untwists upon cooling. Now TS consists of:

(a) thermal field as source of energy, control unit
(b) nitinol cap as engine, transmission, working unit, product

A single component, viz. cap, integrally performs role of as many as three parts of system plus product. TS is leaner, lighter, and less energy consuming than its predecessor; we say system got convoluted from former to latter state. Introduction of a smart material has done the job. *An informal primitive definition of convolution can be framed as below.* When a TS switches to a new state with:

1. reduction of mass, complication, size, power input, losses;
2. perseverance or enhancement of required function(s) performed;

evolution is typed as convolution. Overlapping of several parts of TS is visibly apparent in convoluted systems.

Fig. 1.6 Earlier TS

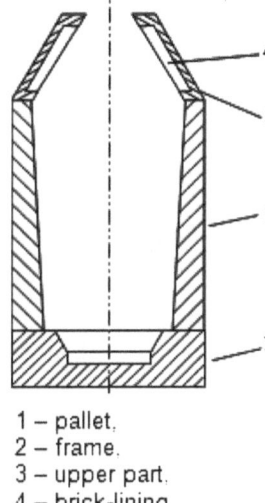

1 – pallet.
2 – frame.
3 – upper part.
4 – brick-lining

2nd Law–Law of Energy Conductivity in System

At the synthesis of TS one should strive for using one field[4] (energy) for all processes of work and control in the system. Every new subsystem added to TS should operate with energy type passing through the system or with free one, i.e. from the environment, waste products of other systems.

Example 1: Use the field that the system contains to solve the problem. At filling up mold with metal, latter can weld onto the brick-lining (casing) or it can flow into the crack between the frame and the upper part. At cooling down, ingot gets smaller in volume, its length decreases. The ingot hovers—it holds on the welded joint between frame and upper part or flows in gap of that weld. At this cracks appear in it under its own weight. How can TS evolve to increase its $Fn(\sum MUF)$; added function to MUF should be getting rid of problem experienced.

Solution: Thermal field is available in given TS. Subsystem added to enhance MUF value must use this field only.

Melted metal is poured. Liquid (water) in inner cavity of piston vaporizes and presses piston. Piston in turn raises gasket. Elevated gasket supports hanging ingot releasing stresses in ingot. If vapor pressure exceeds permissible limit, safety valve acts. If this safety measure is not incorporated, there is danger of gasket and hence ingot overshooting and taking away upper part away from frame by breakage. Part of thermal energy available in TS is diverted to added subsystem. Subsystem uses this energy to heat water. This 'loss' of thermal energy in main TS actually helps us. Ingot cools faster—MUF of TS is raised by another point (Figs. 1.6 and 1.7).

---

[4] Field is used almost synonymously with energy. Finer distinction will gradually emerge later.

Fig. 1.7 Latter TS

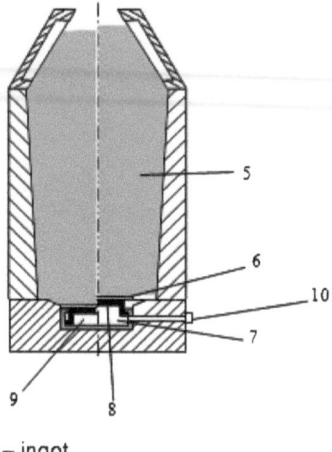

5 – ingot,
6 – gasket,
7 – vapor,
8 – piston,
9 – water,
10 – safety valve

Example 2: The evolution of electro-magnetic relay over years. Electrical and electromagnetic fields were used throughout in advancing TS. Any attempt to grope in a different field, say mechanical, failed.

1st version: Primitive electro-magnetic relay is shown in two positions: switched off as in Fig. 1.8a and switched on as in Fig. 1.8b. Disadvantage: contact is broken at onset of serious vibration. Not robust enough for use in airplanes and spaceships.

2nd version: Electro-magnetic relay with a mechanical latch was designed. Figure for this neither required nor provided. After the switching on, the mobile contact is fixed with a mechanic latch. Disadvantage: to provide movement of latch, separate power supply line, controlling windings, and the scheme of control are required.

3rd version: Electro-magnetic relay with memory was invented Fig. 1.9. Heater is turned on by a separate direct current. Dielectric melts. The contact is turned on. Heater current is reversed—heating changes into cooling. Dielectric solidifies and fixes contact. Disadvantages: switching on takes more time; dielectric interferes with the contact by forming a thin layer between contacts even in closed position.

4th version: Idealization of relay: Soviet Patent 1387069 (Fig. 1.10). Alternating electric current is directed on the winding. Alloy becomes liquid. Alternating current is suddenly replaced by direct current. With this, alloy closes the contacts by stretching along magnetic field lines. Direct current, even though running cannot cause producing eddy currents in alloy; its heating action vanishes. Alloy solidifies. Direct current can be switched off comfortably now—delay in switching it off has no harmful effect except little power consumption. Current of dozens of amperes is controlled under extremity of jerks, vibrations, etc. Notice idealization here?

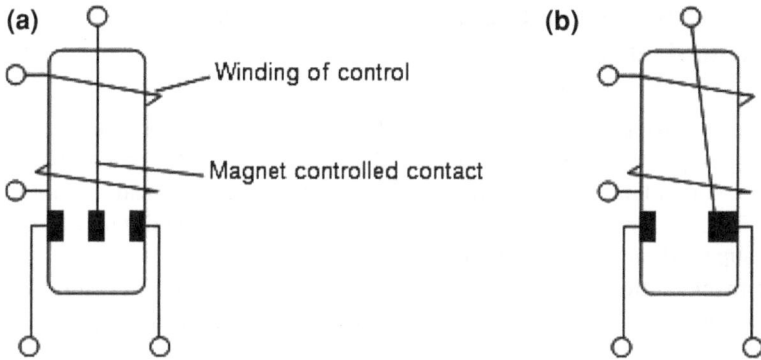

Fig. 1.8 (a and b): an earlier version of relay in two modes

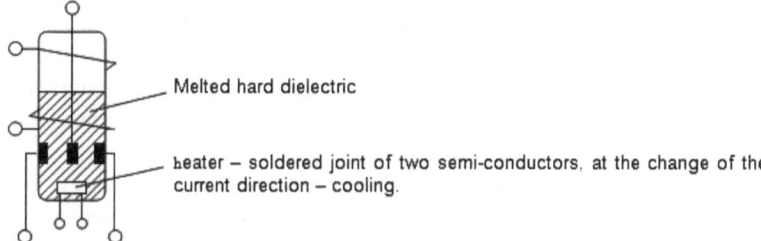

Fig. 1.9 Relay with problems

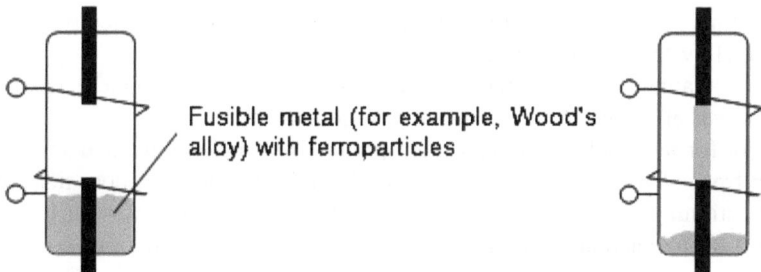

Fig. 1.10 Highly idealized relay

The mobile contact has disappeared; its function has transferred to an alloy. There are no superfluous subsystems and substances like heaters, fusing dielectrics, mechanical catches, etc. MUF has increased, MDE has decreased. Ideality has greatly increased. Convolution of a superior kind has incurred.

3rd Law–Law of Coordinating System Rhythms or Law of Harmonization

In technical systems, the action of the field should be coordinated with proper frequency of product or instrument. The objects sway at the highest amplitude with the exact coincidence of the frequencies. During this harmonization, minimum of energy is spared from outside to maintain resonance, and maximum of the energy that is supplied enters the system.

The reverse effect is also useful in many cases. Prevention or neutralization of resonance is a de-coordination of self-frequency of system with frequency of outer influence or organization of counteraction.

Example 1: A confusing situation appears in the area of ultrasonic metal machining. Though having unique possibilities, this technology is extremely naughty. The basic disadvantage: the generator frequency is tuned into self-frequency of instrument oscillations under free running. Free running implies no load on the instrument. But as the instrument starts working, it is influenced by different tensions. Its frequency changes immediately and stops coinciding with that of the generator; the system is no more in previously set resonant mode. Its coefficient of efficiency rapidly drops (Fig. 1.11).

TS is ultrasonic machine. Metal lump being processed is product. It is very hard to get rid of mismatching in ultrasonic machines under loaded conditions. Instrument's self-frequency is influenced by several factors like alternating (vibration) properties of processed material, the force by which instrument presses half-finished product, conditions of ultrasonic cutting, etc. In order that a substantial part of energy should reach the destination, one has to increase the power of actuators and generators by extremes. But it is utterly ineffective, prodigal method. How to solve the problem efficiently?

IFR is framed. IFR: let the system choose the most profitable (resonance) frequency at all times. Autoresonance is required!

Sensor for feedback, a standard microphone, is set behind the actuator from the side, in front of zone of processing. Microphone in this position does not interfere with functioning of machine. Electrical signal from microphone is transferred to a magnetostrictive actuator winding via an intensifier. Autooscillations appear; their frequency keenly reacts to any changes in working conditions. Continuously adaptive resonance enables effective ultrasonic energy transmission Fig. 1.11.

In process of work (interaction) different parts of the system, mainly instrument, and product should be coordinated to each other at identical frequency for better cooperation or de-coordinated to each other at discordant frequencies to prevent harmful coordination. Moreover, it is profitable to coordinate or de-coordinate not only frequencies but also characteristics like speed, mass, size, form, elasticity, etc. influencing these frequencies. Sometimes the notion of frequency does not even occur in the solutions.

Example 2: When an airplane lands, clouds of smoke can sometimes be observed. Reason: as wheels touch ground, there is a hit resulting in twisting and

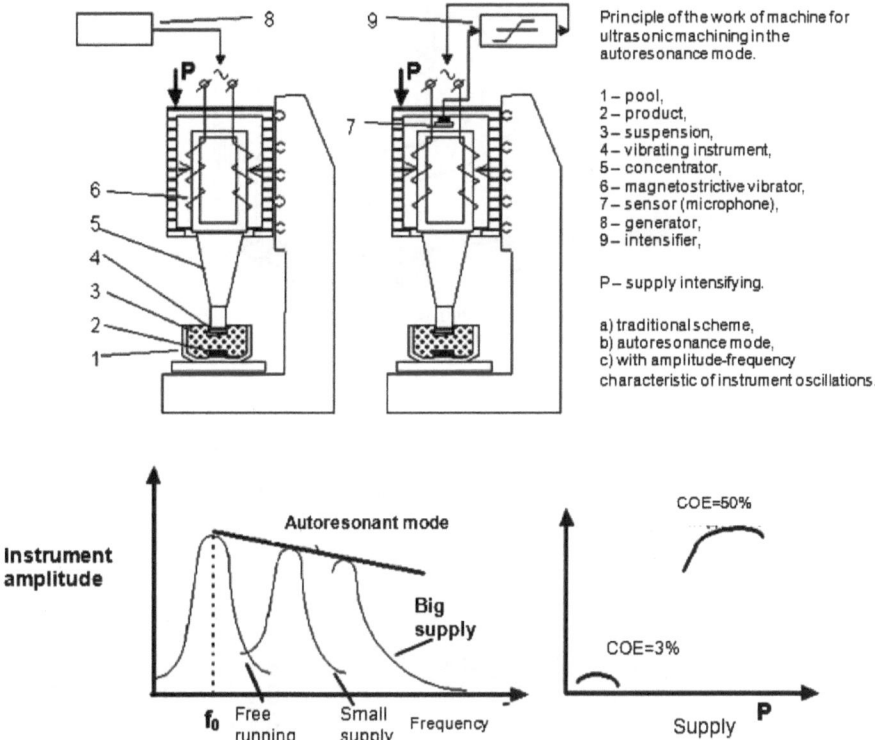

Principle of the work of machine for ultrasonic machining in the autoresonance mode.

1 – pool,
2 – product,
3 – suspension,
4 – vibrating instrument,
5 – concentrator,
6 – magnetostrictive vibrator,
7 – sensor (microphone),
8 – generator,
9 – intensifier,

P – supply intensifying.

a) traditional scheme,
b) autoresonance mode,
c) with amplitude-frequency characteristic of instrument oscillations.

Fig. 1.11 Ultrasonic machining

subsequent slipping of wheels. At this the wheels fray soon. Here is an evident de-coordination between rhythms of the wheel as instrument and leg as product. According to the French Patent 2 600 619, it is suggested to set blades on the side surfaces of the wheels. Counter flow of high velocity air pre-twists the wheels in opposite sense before landing. Net deformation is close to zero (Fig. 1.12).

Problem 3: When wind blows, wires of power line sway. You might have listened to singing wires, whistling pipes, etc. during storms. If flaws coincide with their oscillations, the wires break. What will you suggest?

Multicore wire replaces ordinary set of wires. Outer wire is chosen as thin-shelled hollow cylinder of greater diameter while inner wire is a solid cylinder of lesser diameter. Inner wire(s) is nestled within outer. Dynamically, system behaves as though two springs with different values of elasticity are placed one within another. The natural frequencies of this system are no more discrete values, rather a subsided continuum graph results. Sharp resonances caused by wind are de-coordinated.

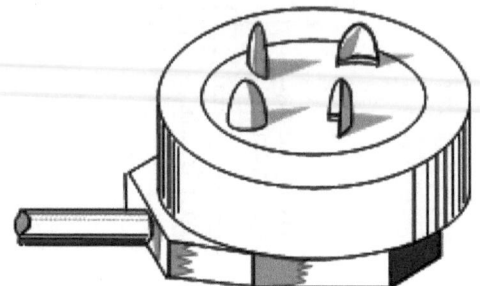

Plane wheel according to
French Patent 2 600 619.

Fig. 1.12 French patent in aviation

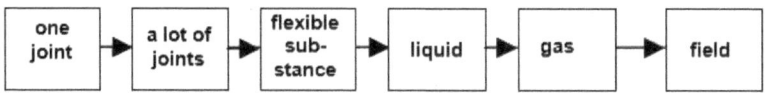

Fig. 1.13 Dynamization of matter

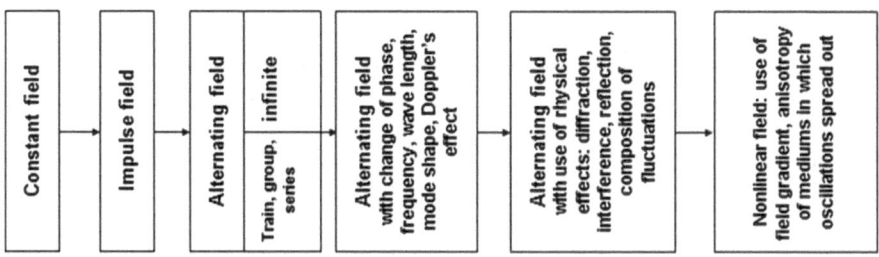

Fig. 1.14 Dynamization of field (energy)

Not a Law but a Strongly Observed Trend: Dynamization

Dynamization of substance: begins usually from the partition of a solid substance into two parts sharing a joint. Then dynamization follows this path: one joint—a lot of joints—flexible substance–liquid–gas. Sometimes dynamization ends by the replacement of gaseous substance by a field tie Fig 1.13.

Dynamization of field—in simplest case is carried out by transition from constant action to pulsed one, then to alternating and nonlinear types Fig. 1.14.

Both chains of dynamization reflect only to most characteristic stages of changes in systems. All stages are not necessarily 'passed through' by systems and not all systems reach the end of chains in their development.

Example 1: What will futuristic cranes be like? Contemporary ones have an ancient winch (monkey engine) added to a brace (arrow). This brace cannot bend over; crane can only take cargo found exactly under rollers, through which a rope

Fig. 1.15 Crane birds

with hook moves. Even most advanced cranes with telescopic arrow and hydraulic control cannot take a weight in the aperture of built building or take cargo from any construction nook. Cranes will have this disadvantage until arrow becomes flexible, like a swan-neck (Fig. 1.15). Such arrow has been invented in Russia. A simplistic version of it is rendered here. Steel disks with decreasing diameter are jointed peripherally by as many as 16 elastic arrows. A flexible carcass shaped like a pipe is generated, giving arrows required stability. Steel ropes, two or four in number also connect all steel disks peripherally—ropes are configured diametrically opposite. If one rope is drawn by hydraulic actuator on the rotating platform situated on operator end, arrow bends in the most fantastical way. Cargo captured by such arrow can be delivered through window in basement of built home or nook of a construction site (Fig. 1.16).

Example 2: Here is an invention of year 1949—the method of producing concave mirrors for reflex telescopes. Process is thus: silver is placed into chamber, cover is welded and chamber is heated by oxydric torch up to the melting point of silver. Chamber is rotated by electric motor. Liquid silver forms an ideal parabolic surface. Torch is shut down. During entire manufacturing, vacuum pump operates to isolate air from silver. This is necessary because air can be absorbed by melting and latter its output can give rise to pores on finished surface.

Problem 4 Microprobe of material from miniature objects, whether unique or cheap, is necessary when a sample has to be collected for immediate or future researches. Usual method is: some site of object is covered by substrate, for example Lavsan[5] film of high purity, which is transparent to rays of laser. The operator turns on pulse laser. Spraying of material on inner (lower) side of substrate occurs (Fig. 1.17). Two disadvantages exist with this method:-

(a) Transparency quickly decreases at the spraying and that is why the probe can be spoilt by laser ray.

---

[5] Lavsan :a polyester developed in Soviet Union.

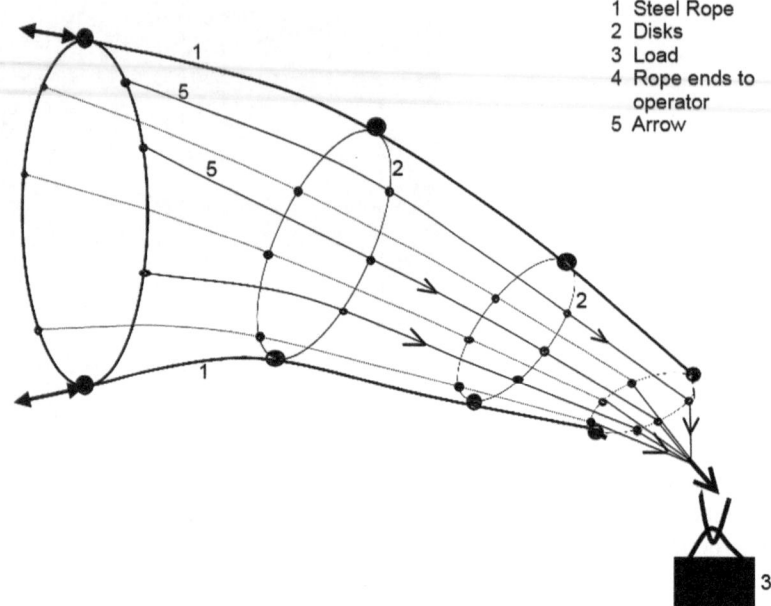

1 Steel Rope
2 Disks
3 Load
4 Rope ends to
  operator
5 Arrow

Fig. 1.16 Next generation crane

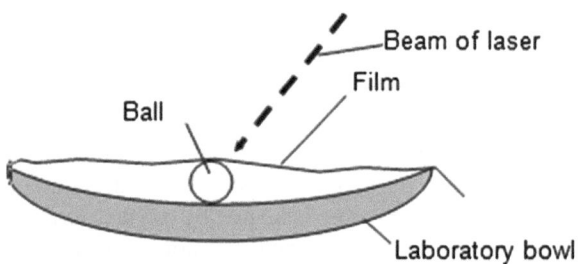

Beam of laser
Film
Ball
Laboratory bowl

Fig. 1.17 Microprobe in attempt

(b) On the one and the same location of film, vaporous minute particles and the liquid drops of bigger size are sprayed. These have to be separated during analysis.

Let us present problem in detail.

1. There is a miniature object.
2. It is necessary to take microprobe of substance from this object. After experiment this substance should be saved.
3. One chosen site of miniature object is covered by transparent film (substrate).
4. There is a laser and the operator turns it on. Beam is directed toward chosen site. Beam penetrates film reaching site.

5. What will happen under action of the laser radiation? The substance of the object will vaporize and settle on the back side of the film.
6. Process can be repeated for different sites.
7. There are several disadvantages in this process:

(a) Transparency of film quickly decreases at the vaporizing. To continue process, laser intensity is raised. This spoils the probe.
(b) Minute particles and the liquid drops of bigger size are settled on the same site of film. They need to be separated during analysis. Solution of one problem causes another problem.

Some scientific history on such sampling:-

After about 100 years, scientists investigated riddle of the Tungus meteorite that fell in Siberian taiga in 1908. All unusual subjects, such as particles of substance, test samples of ground, part of trees, etc. found at that place are stored in several museums of Russia. When analysis commenced, more than 30 hypotheses about nature of this phenomenon existed, but the final solution was missing. To validate or dismiss a hypothesis, there was necessity of a check. Exhibits were taken from museum and researched on. During many tests, black balls, resembling drops of the fused glass of diameter 2–5 mm, were found. All fine methods of substance analysis: x-raying, nuclear magnetic resonance, weight–spectroscopy etc. were at disposal. It was enough that microprobe (micro amount) of substance was made available for these methods. To take such microprobe, earlier mentioned methodology was used. Improvement in this methodology was demanded.

Solution: It is necessary that:-

(a) transparency of a film either does not reduce or does not play any role in method,
(b) particles of different size reach different sites of film (substrate).

It seems ball of sample and laser beam can be kept unchanged while other parts need to be changed.

Evolved methodology: physical effect used: division of particles based on mass when these particles are set into rotation. Dynamization trend comes in effect. Control answer: substrate is established vertically. Sample is rotated with such speed that products of erosion under the action of centrifugal forces get different accelerations and fall on substrate at different levels. Transparency of film is not necessary as laser beam has been spared from system (Fig. 1.18).

Why has effect of gravity been neglected? What will deposit radially outwards: smaller or bigger particles? Answers are left as exercise for readers. As final note to this case study, note that TS has contracted with MUF increased. TS appears leaner, yet more useful. TS has convoluted. Before convolution is introduced formally, we want some informal ideas on it to sink in your minds.

Fig. 1.18 Microprobe in
success

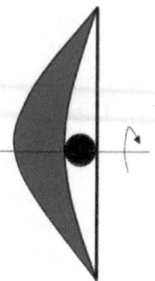

4th Law–Law of Increasing the Degree of Substance-Field Interactions of Technical Systems

A TS is composed of several parts, with each part composed of a single or a set of substances. These substances interact among each other via energy transfers.

TRIZ vocabulary is expanded to proceed comfortably. If substance '*A*' applies a force, say mechanical push *F*, on substance '*B*' resulting in actual or a tendency of displacement, deformation, etc., we say a mechanical field exists between *A* and *B*. Better still we say, Field (mech) acts from *A* onto *B*. *A*, *B*, and *F*(mech) constitute a Substance-Field Model (SFM).

$$A \xrightarrow{F_{mech}} B$$

The development of technical systems goes in the direction of increasing of SFM's degree: non SFM systems aim to become SFM, and in already SFM enabled systems, development goes by means of the increase of the number of ties between elements, increase of sensitivity of elements, and increase of the quantity of elements.

Example 1: See Fig. 11.9. A bushing (part 2) is manufactured by molding. It becomes difficult to remove it from blind hole (part 3) as edges of bushing cannot be grasped. Primitive way: make a flute inside bushing and to tear it out from case of detail by using some 'pulling' tool. Method is time-consuming, unreliable. And for mass disassembling it is totally unacceptable.

Much more effective solution (Fig. 1.19 once again) is offered: oil is filled in aperture, steel roller is inserted, and impact by hammer is made on roller. External action causes hydraulic impact of oil on bushing. With each blow of hammer, roller sinks deeper, causing oil's compression to increase. Hydraulic pressure of oil on bushing builds up. With last blow, bushing pops up. Method is safe. Also notice that pressure of oil on bushing is uniformly scattered on latter. Hence there is no deformation of bushing. Figure 1.20 illustrates SFM of this solution.

Example 2: One more safe method of achieving high pressure in oil is rendered. Earlier version of this technique (not shown): inject nitrogen in chamber, raise pressure by heating. Method susceptible to explosion and very high pressures impossible. Refined method (Fig. 1.21): Chamber is filled up by oil, valve is

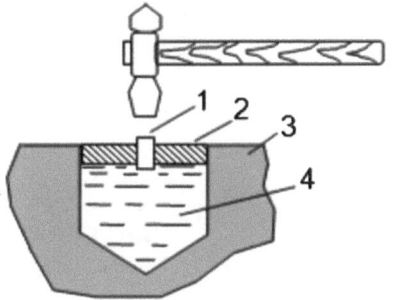

Method of molded
bushing drawing.
1 – roller-piston, 2 – bushing, 3
– body, 4 – oil.

Fig. 1.19 Superior solution through complexes in SFM

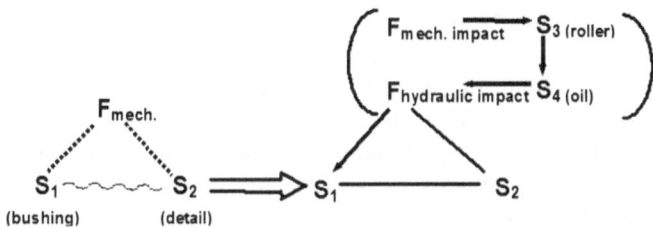

Fig. 1.20 Complexity of SFM is solution in itself

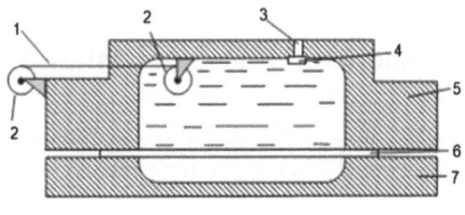

Method of producing of high pressure,
patent 566 656.
1 – Elastic element (belt), 2 – drums, 3 –
aperture, 4 – valve, 5 – chamber, 6 –
заготовка, 7 – mould.

Fig. 1.21 Another mechanism to build great pressure in oil

closed, flexible element (belt) is moved from an external drum to be wound on internal drum. Thus, volume of flexible element inside chamber is increased. Pressure sharply rises. Technique is used in calibration of half-finished material. Readers are left to draw SFM of previous and refined solutions as exercise.

Fig. 1.22 Split connection in
electrical rotating machines

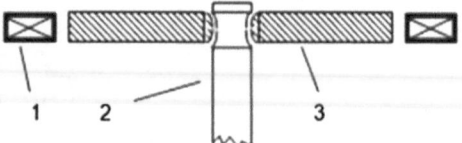

Split connection according to the
Patent 1 298 439
1 – electromagnet, 2 – arbor, 3 – bushing.

Fig. 1.23 SFM of above
split connection

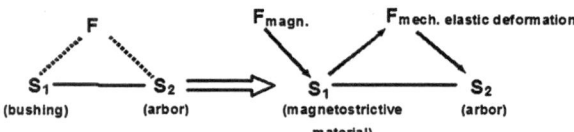

Example 3: Figure 1.22 displays split connection of bushing-arbor-bushing. Evolution: in order to decrease labor-intensiveness and increased productivity of assembling-disassembling of connection, covered bushing is made of magnetostrictive material with positive coefficient of magnetostriction, and its elastic deformation is carried out by an externally influencing electromagnetic field. Figure 1.23 shows corresponding SFM diagram.

In this example, weakly controlled fields like mechanical—setting, beading, thermal—hot setting are replaced by a well controlled magnetic field with simultaneous replacement of substance of the bushing.

5th Law of Transition from Macro- to Micro-Level

As TS evolves, molecules, atoms, ions, electrons, etc., that are easily controlled by fields with the help of physical–chemical effects replace wheels, shafts, gears, etc (Figs. 1.24, 1.25).

Example 1: Evolution of the substance of automobile tires:

(a) tire of solid substance
(b) tire with an air cavity or pneumatic
(c) poly-pneumatic tire: cavity separated by barriers
(d) macro pored tyres
(e) tires of Capillary Porous Material (CPM)
(f) tires with cavity filled with pored polymeric particles, gels, etc.

Example 2: Evolution of lifting methods of sunken ships followed by the way of using:

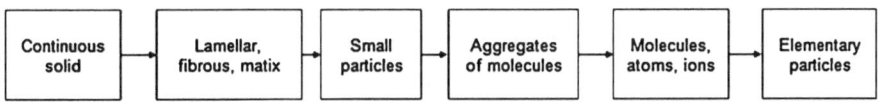

Fig. 1.24 From large to small

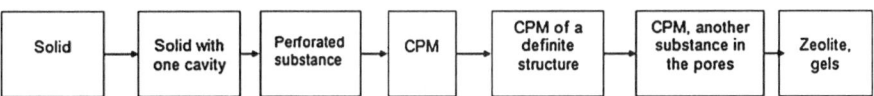

Fig. 1.25 From large to small

(a) Solid body, e.g. lifting of sunken ships with ropes without water removed from internal rooms of the ships.
(b) Large cavities, e.g. pontoons[6].
(c) Plenty of small cavities, e.g. hard foam—FPU.
(d) Grinded foam, e.g. balls, granules of foamed plastic.
(e) Germs of foam, e.g. rubber granules, microcapsules.

Problem 5: Elevated industrial-scale plants, like derricks for petroleum and gas, require constant presence of manpower. It can be sometimes be very dangerous for people. In an emergency there can be emissions and flashes of fuel. The problem of fast evacuation of people is very important. Usual multilevel ladders are absolutely useless; people have no time to go down to reach ground. To minimize risk, special lifts which move in a steel pipe near to the industrial-scale plant, were installed. On top platform, people run for lift, close the door, and press the button. But even high-speed lifts always cannot provide desired safety since unwound ropes have inertness. Besides during accident, power cut is likely disabling lift.

The following solution would be ideal. Electricity is not required. Cabin of the lift freely falls down, but does not break. It smoothly lands after which people run out from cabin to safer places. What is your suggestion?

Solution: You should take into account following:

(a) It is the height of 30–40 m, i.e. equivalent to 10–12 storied building.
(b) It is difficult to imagine a material: absolutely inflammable, easy, thin, with a high degree of thermal protection.
(c) the sleeve should be in a twisted condition normally and should open at the fire. This mechanism must be reliable, especially unaffected by likely winds.
(d) most important: in such cases there is a panic, any slogans and suggestions will not help.

---

[6] But for Russian deep sea vessel 'Peace', grand film Titanic could not have been produced.

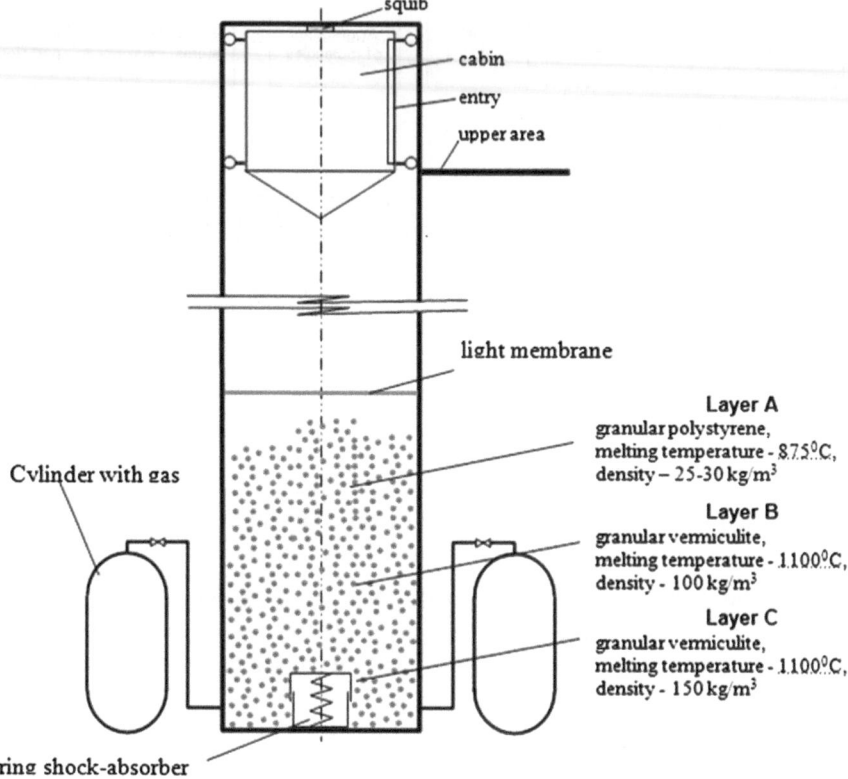

Layer A
granular polystyrene,
melting temperature - 875°C,
density – 25-30 kg/m³

Layer B
granular vermiculite,
melting temperature - 1100°C,
density - 100 kg/m³

Layer C
granular vermiculite,
melting temperature - 1100°C,
density - 150 kg/m³

Fig. 1.26 An extraordinary life-saving patent

(e) fight near the sleeve can occur. It is necessary to provide easy access for all
people simultaneously in a saving cabin (capsule) and instant disappearance
from the place of accident.

Solution: Patent 128789 Fig. 1.26. People enter into a cabin, a door tightly
closes, people sit in armchairs, squib operates, cabin falls downwards, valves of
cylinders with gas open, a boiling layer of granules is formed, the membrane
breaks, the stream of granules flies toward cabin smoothly stopping it. The part of
granules of polystyrene move onto roof of cabin and if there is fire they melt and in
addition protect cabin. Notice how smart substances are better than components/
subsystems of components in a wonderful convolution like this.

Chapter 2
Origins of Convolution

Law of transition to the super-system subtly adds positivity to TS while deducting negativity simultaneously. System gets more suited for required application, yet does not complicate. This law is better understood through extended examples. Convolution gently enters life of TS.

When and who invented to put a slice of cheese, sausages, or something else on butter? Sandwich: two slices of bread with butter, between which there is a chunk of meat, was invented in eighteenth century by English admiral Lord Sandwich. Such sandwich was more convenient to eat during card game; butter did not dirty cards. Can we call a sandwich a bi-system composed of two mono-systems, two slices of bread, plus something extra?

Aperture of the camera was installed before or behind lens. In the former arrangement, image became a little swollen, straight lines turned into convex. In second case, image became a bit shrunk-straight lines turned into concave. This phenomenon, a kind of optical distortion, could not be eliminated for a long time. The solution was finally found: to set two apertures, one before and other behind lens. The beam first expanded, and then shrank almost identically. Distortions were mutually compensated. It is a classic case of minus and minus becoming plus in algebra. System quality appears in Bi-system from two similar elements, in case if they fulfil the counter (conflicting, contrarily directed) functions. Examples: pair of scissors, apertures in camera. System quality also appears in Bi-system from two differed elements (with shifted characteristics) like in famous twin-blade of 'Gillette'—one blade raises hair, second one cuts it.

Look at a usual drill. Guess in which way will it evolve? We attempt inventing a new drill which is better than known. Conventionalists might offer faster production of drill or more sharpening to it. In such offers, the system stays as system: psychological inertia does not let system pass on to bi-system or to poly-system. Right answer is: drill must become binary drill. What does it means to connect two drills? It is necessary to take two drills and to turn their in one binary drill. Like a double barrelled gun with two barrels sharing single buttstock or a

S. Kwatra and Y. Salamatov, *Trimming, Miniaturization and Ideality via Convolution Technique of TRIZ*, SpringerBriefs in Applied Sciences and Technology, DOI: 10.1007/978-81-322-0737-5_2, © The Author(s) 2013

two-color pencil–one end blue lead, another red lead, sharing common outer wooden cylinder. Binary drill has spiral rifling at both ends. When one end gets blunted, it can be turned around and the worker can work by second one.

Transition to bi-system and poly-system is an inevitable stage in the development of every system. For example, ancient anchor was a hook with one fluke. Then anchors with two flukes and polyflukes appeared. A drawing-pin (thumbtack) with one spike is simple system. But bi-thumbtack (two spikes) and poly-thumbtack (three spikes) were invented. Such transition increases MUF of system.

Nail is a simple single system. What will be its form if it passes onto a poly-system? Poly-nail was developed by Finnish engineers. It has a metal plate with multitude of spikes. One head is enough for 200 spikes. Wooden constructions using poly-nail is two times faster than usual. Take notice: it is incorrect and no good to simply add systems together. One cupboard stacked on top of another serves no extra purpose; it is, if at all, a low-grade invention. Addition must give benefit; bi-system should be easier or efficient than two separate systems considered together. When two guns are joined to form a double barrel gun, several parts are eliminated. The most advantageous trick is to unite something with nothing, emptiness, free of charge resources. To prevent injury during training dives, bottom layer of water in swimming pool is united with air; air bubbles are released and diffused from floor of pool. In comparison with dense water, air is almost emptiness. Together they form a 'soft' layer of water at bottom of pool, producing a cushion effect.

Example 1: Magnetostrictive pump: A pump with magnetostrictive elements powered by ultrasonic generator is invented. Patent 885635. Fig. 2.1.

Element 1st and 2nd are magnetostrictive working elements. If an element operates singly, it is a mono-system. It can just achieve shaking of liquid i.e. it pushes it forward–backward. Maybe in some consideration this is a useful function. When we join these two similar elements with their phases shifted, it is bi-system with biased characteristics. A novel system property, to pump liquid, is gained. Here the positive properties of elements, viz. to push liquid forward are added while negative properties, viz. to push liquid back are eliminated. This bi-system fulfils two functions:

(a) pumping function, transferring of liquid forward,
(b) valve function, locking the motion of liquid backward.

Both functions are actually part of single MUF of this TS—pumping liquid. It is mentionable that treatment of step-by-step working of this system is beyond context of this book.

Example 2: Welding. We begin with Laser Beam Welding frequently referred to as LBW. In this process, multiple pieces of metals are heated to a molten state and fused together using lasers. TS consist of two or more portions of metals to be jointed, laser beam along with its subsystem for optical focussing feedback and filler wire. TS is a mono-system so far. Some of you may disagree, arguing that two or more metal parts are involved. Our answer is: let the mono-, bi- and poly-nature of TS be set according to numerical value of laser beams.

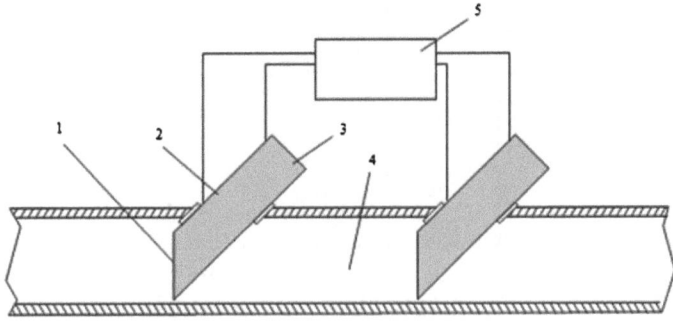

1 - section of the perpendicular axis of pipe,
2 - the end of working element,
3 - magnetostrictive working element,
4 - working chamber of pump,
5 - ultrasonic generator

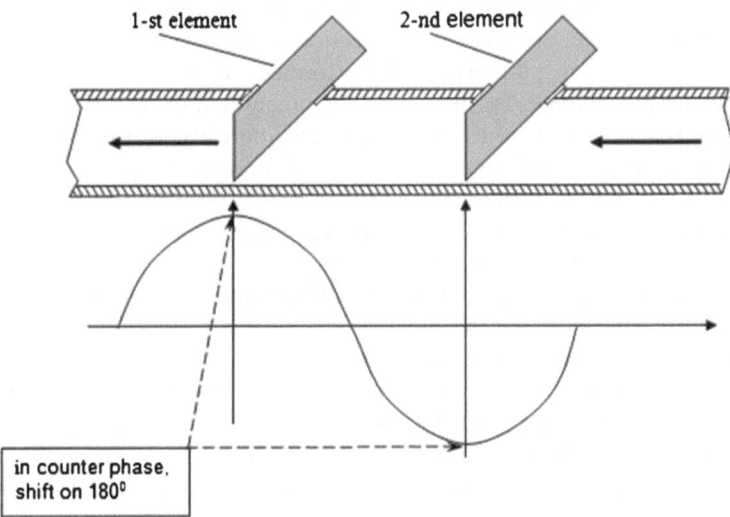

Fig. 2.1 Bi-system with biasing between parts manifested as phase lag

To increase MUF, mono-system is progressed toward bi- and poly-systems. Few possible directions are given herein:-

(a) Either the beam is divided into two beams, or two separate beams are used. The two beams are unequal in intensity and/or frequency. They are focused symmetrically about direction of welding. This is bi-system with heterogeneous characteristics. New system property: in case of dissimilar metals

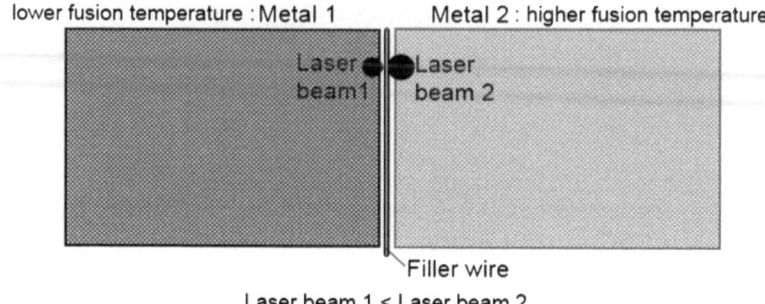

Fig. 2.2 Different metals require different heats of fusion

being welded, different thermal powers are required to melt each of them. Dissimilarity in lasers' power matches this, producing a homogeneous weld Fig. 2.2.

(b) In case (a) above, if identical metals are used, laser beams may be identical. TS is now a bi-system with homogeneous characteristics.

(c) The two laser foci are aligned in direction of welding with one leading the other. New system property: keyhole formed due to alloy vaporization is stable, assumedly because the keyhole is bigger and the evacuation of metal vapors is eased. Consequently there is less porosity within the weld, and blowholes (local explosions of the weld) almost disappear. Mechanical strength of weld is greater. In long run, usage of superiorly welded metal reduces scrap and repair of welded components. This is bi-system with heterogeneous characteristics (Fig. 2.3).

(d) Galvanized steel, viz. steel coated by zinc on both sides, is currently used in automotive industry for the manufacture of car bodies, especially for its panels and structure. To produce thicker layers, galvanized steel is welded sheet on sheet with resistance spot welding. In doing so, the zinc layer is vaporized easily because of the high current used, and the high amount of heat produced at the sheet/sheet interface results in a good weld. When laser welding of galvanized steel was introduced, engineers faced a problem: vaporization of zinc perturbed the keyhole, result of which several defects like porosity were introduced. To proceed, laser beam was divided into two parts: the first advancing part was not focused on the material, so that only conduction heat was produced, but in sufficient amount so that zinc vaporized. The second part produced a keyhole for welding. New property: as zinc is already evaporated when the welding occurs, the weld is of excellent quality. Thus galvanized steel is now 'laser weldable'. This is bi-system with biased characteristics.

(e) Laser welding of tailored blanks has becoming usual for the steel companies which sell their products to the automotive industry. Tailored blanks are usually composed of two sheets of different thicknesses which are butt welded. Corus company has developed a method of producing aluminum

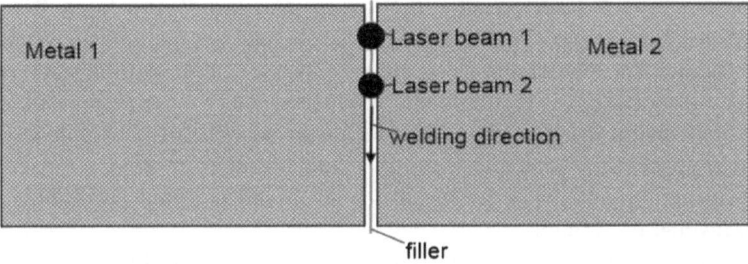

Fig. 2.3 Two laser beams following one another

tailored blanks at industrial scale for Lamborghini Gallardo. For a long time, aluminum tailor blanking was dissatisfactory due to poor quality of the laser weld root: spikes were produced because of keyhole oscillations, there was lack of fusion. Spikes are problematic for further forming (stamping), and lack of fusion may lead to forming failure or, what is worse, to a failure in service because of fatigue soliciting. In order to solve that problem, the laser beam has gone to a bi-system process: on one side, a laser beam classically welds the two aluminum blanks in butt configuration; just afterwards, a few millimetres away, on the other side, a non-focused, second laser beam (diode laser) re-melts the root of the weld or maintains further the melting of the weld root, smoothing the root weld. New property: at the weld root, the spikes are eliminated; fusion is complete. Consequently the high quality of the weld root eases the forming of the tailored blank. High mechanical resistance of the part is ensured. This is a bi-system with more heterogeneity.

(f) MIG-welding: MIG-welding is interesting because it brings a large weld, but the depth of the weld and the welding speed are very limited. Laser welding is interesting because the welding speed is high (good for industrial productivity), and the weld is deep, but the quality of the weld is not as good as that of MIG-welding, and the width of the weld is narrow. MIG and laser welding have been put in synergy into one single welding process, called hybrid welding. New property: the process cumulates the benefits and erases the drawbacks of each process because its speed is high, the weld is of excellent quality, and its width and depth are high. This is also a heterogeneous bi-system.

(g) Friction stir welding (FSW) is a solid-state welding process, which allows the welding of any aluminum alloy including so-called non-weldable alloys by fusion. The drawback of this process is its low speed. By adding a non-focused laser beam in front of the FSW tool, the material is heated, its plastic flow stress lowers, therefore easing the advancing of the tool. New property: the FSW tool can go at a higher speed; this increases productivity. Once again a heterogeneous bi-system.

(h) Laser with Anti-laser: What is the inverse of a laser? Anti-laser. What is its property? When an anti-laser crosses laser, both should vanish. What could be the interest to couple a laser and an anti-laser, with still the same purpose to weld metals? If it is difficult to produce a pulsed laser beam with exactly desired characteristics, it is possible to do so optically: an anti-laser with some space–time characteristics is made to cut a continuous laser beam. Result: continuous laser beam turns pulsed with required properties. This is an inverse bi-system.

(i) Welding conjunct its inverse: What is obvious inverse function of welding (joining)? Cutting. Indeed lasers are also used in the industry to cut metals. Actually CO_2 lasers can do both: welding and cutting. One can imagine the following: the same high power source laser is used at the same time for cutting and welding. How to do so? The laser ray is divided into multiple laser rays by the means of semi-reflecting mirrors. The laser cuts the blanks of different gauges and welds later the different blanks into so-called tailored blanks, with different parts of itself modulated in energy according to task to be done. New property: higher laser ray stability; enhanced productivity. New properties possible: one may imagine a laser that performs multi-tasks with a high degree of quality: welding, cutting, shock preening (for better fatigue performance), printing, etc. This is a partially convoluted bi-system.

(j) Mono-substance: We have seen that for the stability of the capillary (keyhole) during the welding of aluminum alloys, it is necessary to use two laser beams. Can this stability be achieved by a single laser beam if it is coordinated with substance being welded? It has been shown that the use of one single laser beam stabilizes keyhole, if the beam is pulsed with some definite frequency, and with time gaps in some multiple of wave period. It is if these parameters are coordinated with time-driven functions of the keyhole physical phenomena, optimized result is obtained. Law of Harmonization, our third law of technical systems' evolution. New properties: better keyhole stability, and consequently less porosity and explosions (blowholes). We can call this a mono-system with monosubstance.

(k) Multiple laser beams' welding: One may easily imagine putting together three, four identical pulsed laser beams together, so that the stability of the keyhole during welding is much more stabilized. The laser beams could be also of different energies, different shapes, etc. Other functions may be also realized. Convoluted poly-system.

Example 3: Boat. Two boats coupled by transverse beams constitute a homogeneous bi-system called catamaran. New property: horizontal stability allows higher and larger sail, so that the wind driving force is higher, and consequently the catamaran is speedier. A new TRIZ term introduced here: the joining, internally running substance is called internal medium. In this case, transverse beams constitute internal medium.

Three boats transversely connected constitute a homogeneous poly-system called trimaran. It possesses one central body and two lateral floats.

New property: same as in catamaran, plus its performances are higher than those of catamaran. For that reason, trimarans are much liked for in sailing competitions. Internal medium: beams between body and floats.

A trimaran hydrofoil with water ballasts in the lateral floats constitutes a poly-system with biased characteristics. Usually ballast is used for submarines. In our case, ballast on the side of the wind is full as while on the other one is empty. New property: Enhanced stability that allows easier hydrofoil effect.

Chapter 3
Idealization and Convolution: Two Sides of Same Coin?

Emergence of Convolution in TS Evolution: Concise Treatment

1. Every now and then, man finds deficiency in TS created by him. Requirement to increase its MUF emerges. Time and MUF move in one and the same direction—forward. This permits both time and MUF to be marked on a single axis, generally x-axis. Of course former progresses continuously, while latter discretely. Every step taken by TS marks a jump in MUF—signifying that an invention has taken place regarding that TS.
2. To increase MUF, one has to strengthen (accentuate) a relevant property of one element of system. Efforts are thus concentrated on this part or element of TS. *Evolution of the systems goes irregularly: the more complicated the system, the more irregularly the development of its parts goes.* This is the seventh Law of Technical Systems' Evolution.
3. Several mechanisms are available to advance this 'chosen' element—one of them being differentiating it into several zones with specialized property or chosen value of property ascribed to one zone. This reminds us of beginning of engineering itself; allocation of working organ, transmission, energy source, etc. from mono structure took place.
4. At accentuation of one property of one 'favoured' element, coordination with other elements of TS is broken. A technical contradiction (TC) arises. This step is elucidated with examples.

   Example 1: Cellular device is TS here. Increase of MUF comprises raising capabilities of TS, loading it with modern features like multimedia, camera, bluetooth, Wi-fi and list is endless. TS consist of two chief elements: power block and electronic block. Generally, former is battery and latter is cell phone without battery. The favored element is electronic block, favored more by the electronic and communications revolution than by anything else. With every

S. Kwatra and Y. Salamatov, *Trimming, Miniaturization and Ideality via Convolution Technique of TRIZ*, SpringerBriefs in Applied Sciences and Technology, DOI: 10.1007/978-81-322-0737-5_3, © The Author(s) 2013

new feature added to cell phone, MUF advances. To achieve this, electronic block advances spectacularly. It becomes more sophisticated, yet smaller in size. To get an idea of size, compare gas filled diode valves with today's integrated chips of nano dimensions. But the power block is neglected more because of fact that material science and chemistry cannot keep pace with IT revolution of exponential nature. A contradiction arises:-

If electronic block is advanced, power block has to become bulky. We have powerful (hence bigger) batteries for smart handhelds like BlackBerry. The size reduction achieved by advancement of electronic block is more than compensated by size increase of power block. On the whole, size of instrument increases. Framing it as technical contradiction (TC) in vogue with TRIZ:-

TC—if one property of system, namely ability to offer advanced interface, multimedia, camera, etc. is upgraded, another property of system, namely size, degrades. Degradation of size consists of its actual increase.

Example 2: Tungsten filament bulbs, their mass production, and widespread usage. TS is composed largely of two elements: electric bulbs and manufacturing of electric bulbs. By the beginning of this century, disadvantages of lamps with carbon filament were defined. The carbon filament got destroyed soon, limiting the temperature of incandescence, and the brightness of luminescence. It appeared that filament of some refractory material was required. A. Lodygin managed to make a filament of tungsten and demonstrate such an electro-lamp in the World exhibition in Paris in 1900. Nevertheless, metallurgist could not create a technology of producing thin tungsten filaments. In Europe, there was patented and put into practice the technology of producing filaments of other refractory material—tantalum. Ample production of tantalum was organized. But no other material could compete with tungsten in qualities like fastness, endurance. Hence tungsten was preferred. Tungsten bulbs were used sparingly— they were produced by expensive and delicate craftsmanship. MUF arises now—widespread lighting of homes, offices, and streets via tungsten filament bulbs. Note carefully that tungsten filament has been invented before this MUF arises. The element of TS which is favored is electric bulbs (existence, usage) while element of TS that is neglected is manufacturability of electric bulbs. A contradiction emerges. If tungsten bulbs are used more often, efficient, reliable, and powerful lightning can be enabled everywhere, but production to meet this demand is massive, expensive, and sluggish. TC can be framed here as:

If set of these system properties—power, efficiency, reliability, widespread lighting, etc. are upgraded, set of these system properties—cost of production, bulk of industrial processes, etc. are degraded

5. Resolution of TC is done. MUF gain as proposed is achieved. TS become TS' so to say. Since very soon man becomes dissatisfied with TS', he will propose a new MUF. When that MUF is achieved, TS' becomes TS''. Process is unending. It seems TS, TS', and TS'' are just 'balance points' in ever-refining graph of TS.

Resolution of TC occurs in two distinct ways:

(a) Expansion of TS: Contradiction is resolved in obvious sense. TS' has more MDE than TS. Efforts are made so that MDE gain is not as much as previously speculated—to this end, some functionally useful subsystems are added, some not so functionally useful subsystems are subtracted. But still the net result is the rise of MDE. TC arisen in Example 1, viz. cellular phone is resolved by expansion of TS. The downsized yet smarter electronic block is clubbed or coupled to the larger (may be largeness subdued a little) battery resulting in smarter but bulkier instrument.

(b) Convolution of TS: Contradiction is tactfully resolved by disappearing of subsystems or even TS itself. How can this happen? Either, their functions are transferred to the adjoining systems far away from TS or they are replaced by an ideal substance. This ideal substance is 'smart'—it is programmed to fulfill the function that was previously fulfilled by subsystem or TS. TC arisen in *Example* 2, viz. Tungsten filament bulbs, their mass production and widespread usage was resolved by convolution of TS. Finally, the technology of producing filaments of tungsten cheaply and in huge numbers was worked out and until today tungsten filament electro lamp industry is on run. MDE of TS has fallen in per bulb sense; cost to manufacture, and use of single bulb has dramatically reduced.

A hypothetical idea: If TC was resolved by Expansion of TS, what could be state of TS? We could have an auditorium lit by 100 bulbs alongside 10 factories to produce those bulbs! The compact treatment is concluded by an astonishing revelation. A TS resolves its consecutive TC by choosing expansion mode for few times, followed by convolution for few times, again followed by expansion for few times and so on.

TS $\longrightarrow$	TS' $\longrightarrow$	TS'' $\longrightarrow$	TS''' $\longrightarrow$	TS'''' $\longrightarrow$	TS''''' $\longrightarrow$	TS''''''
resolving TC via expansion	*resolving TC' via expansion*	*resolving TC'' via expansion*	*resolving TC'' via convolution*	*resolving TC''' via convolution*	*resolving TC'''' via convolution*	

Physical reason for TS following this behavior: initially TS uses all available resources, both within and outside (environment) to resolve its facing TC, and increase its MUF. These resources can be substances as well as fields. As a result, MDE rise and TS expand. After few spurts of expansive growth, TS is unable to withdraw from available resources—they have been exhausted. The only choice for TS to proceed is by becoming clever. It convolutes so to say; in doing this it sometimes uses 'hidden' properties of substances composing it, sometimes replaces common materials by smarter ones containing natural programming, sometimes shedding its extraneous subsystems off its sleeve.

Thus, a TS follows a wave of evolution, first half wave of expansion, next half wave of convolution. Of course, this is an unending waveform.

Emergence of Convolution in TS Evolution: Extended Treatment

Analysis of history of all TS shows that all of them develop through a chain of following successive events:

1. Appearance of requirement.
2. Formulation of MUF—social demand for new TS.
3. Synthesis of a new TS, start of its functioning (minimal MUF).
4. Increase of MUF—an attempt to get from the system more than it can really give.
5. At increase of MUF some part or property of TS worsens—technical contradiction appears. Opportunity to formulate an inventive problem.
6. Formulation of the required changes of TS by answers to questions like these: what should be done to increase MUF? And what does not let one do it? Transition to the inventive problem.
7. Solution of the inventive problem using knowledge of science and engineering and at times even culture. It is here that diversification occurs—system follows either expansion or contraction. Contraction is given a special name, convolution. This step is akin to step (5) of limited version above.
8. Change of TS post-incorporation of invention. TS modified to TS'.
9. Increase of MUF.

These 9 separate stages of TS evolution are examined thoroughly. Points are abolished; an integrated discussion interspersed with widespread examples is provided.

Everything created in the world of engineering is made *only* to satisfy the requirements of man and his society. If there is no need for TS, it will never appear. If a requirement appears, then as time flies, requirement becomes sharper and obsessive—nothing can then prevent society from creating a TS and putting it into practice. Of course, society employs innovators for labor. Summarizing this paragraph, requirement is the mother of all inventions.

Post-industrial revolution, expense of muscular power in industry is extremely low, about 0.1 % of total mechanical energy spent in industry. It means that without machines, we can get only 1/1000th volume of current production. Machine power has become significant and will increasingly become in future as remaining labor functions are transferred to technology. Can we call this trend, automation? Need to economize this machine power ought to and has rightly become a vital human need. Of course man has already fulfilled prior needs like food, water, sleep, offspring production, and defense from exterior dangers. These needs in particular provokes inventiveness in engineering. How? This need gets escalated to necessity. This is the beginning of contradiction of individual or society with the environment, breach of the required balance with it. The appeared contradiction becomes a motivation for active work directed to satisfy the requirements. Further technological evolution is impelled, rather compelled.

Progress of the society would have been impossible without stimulating role of requirements. Law of eminence of requirements is a cross-disciplinary law acting on human history independently and objectively, through its subtle mechanism of influence on human will and consciousness. This law has unabatedly governed us from prehistoric ages till today—and there are no signs that it will lose it hold in centuries ahead. Summarizing till now, necessity of satisfaction of constantly growing requirements of the society comes to contradiction with the existing methods of its satisfaction. This contradiction is resolved through power of creative abilities of human mind.

Impossibility to meet increased requirements with old (available) TS makes one invent new system or improve old one introducing new subsystems. The first traffic light, for example, appeared in London in 1868, when intensiveness of movement of carriages exceeded all safe limits. A sharp requirement, movement controller, appeared paving way for creation of a new TS. At the extremely lively square in front of the English parliament, they set poles with gas lamps, which were hand controlled and showed two lights, red and green, through colored glasses. Nevertheless, introduction of a new TS provoked a harmful effect too—the lamps sparked and hissed, frightening horses. It was only in the beginning of twentieth century that traffic lights with electric lamps appeared in USA. It was of horizontal type, with three light filters—red, yellow, and green. The construction turned out to be suitable and soon international standard on vertical traffic lights was accepted.

First aircraft disaster in the world occurred in 1908 for breakage of a mere screw. It was the time when failure of any element on the plane led to flying accident. To increase safety of flying, new ideas of increasing steadiness and control of airplane in the restless atmosphere were in asking. In 1914, in one of the competitions on the safety of aviation, a new airplane equipped with stabilizer of fly speed was displayed. This plane stood the tests greatly during flight from Versailles-Shartr and back at a speed 75 km/h under wind with speed of 15 m/s. Thus requirement of increasing MUF was met with creation of a new subsystem—stabilizer.

The bigger the jerk in MUF is demanded, the harder it is achieved. It is obvious. Often first TS with high value of MUF are awkward, their functioning is on the verge of possible break—but people did this, for example, at war time. Victory depended on lead in inventive race in all spheres of military engineering. In 1943, above Moscow at altitude of 13,000 m (immense for those times) aircraft bombers would appear. For a long time, it had no punishment, as anti-aircraft fire could not reach them and Soviets still had no aircraft that could fly at such altitude. Immediately, a special interceptor was created. It was a category of plane with two specialities: additional air supercharger and plane maximally lightened in order to reach altitude of 14,000 m. Second achievement, viz. dramatic weight reduction was achieved by measures like replacing ironclad back of the pilot chair by veneer one and eliminating all armament onboard except machine gun. Finally the two planes, viz. enemy bomber and Soviet interceptor met at altitude of 13,000 m. A fight was required. But none of them could start fight as they worked at the limit of their abilities. More fascination: enemy aircraft bombers as it turned out did not even have any armament. And the Russian interceptor could not take the position to attack. Both airplanes could hardly make any turn even at large bending radii. Final outcome: having circled, they separated never to meet ever again. Modifications of TS with increased MUF and new helpful subsystems met—but even then a harder jerk in gain of MUF was required. This jerk could not be matched by innovations in those times.

First refrigerator was created by a butter seller T. Moor (US Patent 1803). He distributed his products all over Washington and requirement of such invention were really sharp for him. It was a big box with doubled walls, between which ice was put. The useful function was gained, but ice was prepared in winter and after that it had to be saved, carried, cut, etc. In 1868, a fridge compressor was created to get artificial ice for food stores, chocolates units, etc. By end of nineteenth century, first domestic ice producing machines appeared in market. One of them called 'Eskimo' was sold in Russia. These machines consumed much fuel—firewood, coal and kerosene. In 1911, giant firm 'General Electric' started production of fridges of modern type: this compressor-based machine could be installed inside kitchen cupboard. This fridge was actually invented by a teacher in the monastic college in France T. Oddifent. Compressor with driving belts produced much noise, gasses such ammonia and anhydride sulfide with unpleasant smell escaped to fill the kitchen. In 1926, a Danish engineer A. Stindrup took a further step: he hid compressor with belts under hermetic cowl with isolation. The fridge became noiseless and smell disappeared. First domestic fridge without compressor was based on absorption. It was invented in Sweden by B. Platen and K. Moonters in 1922. Since that time, desperate competitive fight of two types of fridges, with and without compressor, has taken place. In 1951, in the Institute of Semiconductors, Russian Academy of Sciences, thermoelectric fridge in the world was created. Nevertheless, compressor fridges have quickly improved: poly-functional automatic systems appeared that prepared ice from water themselves, drinks were cooled to definite temperature, butter to definite maintained softness, block for forecasting appearance of faultinesses was added, etc.

Time recorders popularly called clocks took a different development curve. Clock as TS with clear and exact helpful function—time count, have experienced long evolution. As the basis of principles of action of this system, one or other periodical processes were relied upon: Earth's rotation in solar clock; pendulum swaying in mechanical and electromagnetic clock; tuning forks in tuning fork clock; quartz plates in quartz clock. Modern electronic watches have a very high value of MUF—inaccuracy of time does not exceed 1 s in 1 year. Why were timekeepers evolved to this degree of precision? Was there a genuine need? No. Degree of usefulness, viz. measuring time to this fineness exceeded requirement, i.e. need for such fineness. This gap had to be closed. So evolution decided to go other way. Needs for precise time were escalated. Results are visible—pressure sensor, pulse recorder, digital temperature, skin resistance measurer in lie detectors, sound and light signalization, diary, notebook, disk players, wireless, television set, games, computer, stop watches in cricket, messaging code signals like 'ambulance' are some examples relying on nano-seconds. Clocks without apparent power supply like self charging from wrist motions and those obtaining energy from the environment were mass-produced. After 90 s watches were absorbed by cellular phones; they automatically set to exact time through network towers. In 2000 s, clocks are embedded in operating systems of computers; they automatically update their time from Internet. Invention of systems with MUF exceeding level of contemporary requirements is not a rare case in the history of engineering. When a wide gap appears, either search for sphere of application appears—one of the problems of marketing or stimulation of requirements starts—extreme advertising, 'upbringing' of consumer. Veritable requirements of society should be distinguished from forced, artificial, and even silly ones. According to the point of view of a famous American sociologist, about 80 % of all produced or sold goods in USA today the fail to correspond to the real requirements or are useless for society.

MUF of systems constantly increases. Recession, hitches, and short pauses occur only when TS comes close to the moment of exhaustion of resources corresponding to the physical principle laid on the basis of the given system. Change of principle of functioning opens new resources for development. A. A. Mikulin, a famous constructor of aviation engines, once said, 'Let's show table of records, starting from year 1904 and everyone will see: steady increase till 1943. And then every new 10 km/h was obtained with great effort and at 700–950 km/h the curve stopped, deadlock! 'I'll explain', went Mikulin on, smiling to explain why it happened, 'The plane faced the sound barrier. Further increase of speed required increase of traction, power in the geometrical progression. And increase of traction is enlargement of dimensions of engine and the whole plane...at this time everyone recollected reactive engines'.

Increase of productivity of computer is followed almost a linear law till about 2002. It is well known that last 10 years has seen 20–30-fold growth, obviously of nonlinear types Fig. 3.1. Forecasters, from non-TRIZ domain predict stall in further rise of MUF after 2020 unless TS shifts to a different level altogether, like molecular, ionic, DNA, pure field etc.

We return to 'normal' TS wherein requirement of MUF is ahead of available MUF. This mismatch, which is generally an inventive problem, is processed till a bold TC is disclosed. Complete resolution or partial resolution (some relaxation) of this TC is done. Result is inventive solution to inventive problem that was faced. This inventive solution helps TS step aside from action of harmful factors that had prevented MUF from increase. New properties, functions appear in the system; substances and subsystems of TS change—transformed or replaced. On the other hand, if compromise is attempted instead of solution, system is left unchangeable on the whole.

Just to let you know your milestone, we are now in step with point (5) of limited explanation and point (7) of elaborate version.

A fierce competition and interconversion between 'substance' and 'subsystem/ system' begins here. To ensure you are firm footed, let us take example of a common and universally likeable TS—motor car. Transmission, engine, suspension

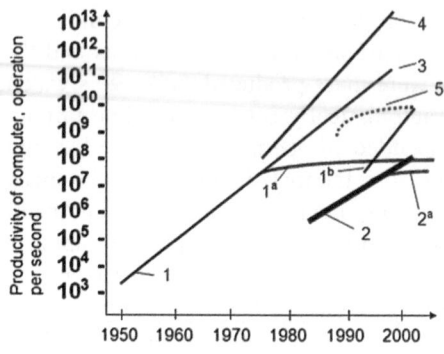

Productivity grow of different classes computers.

1 – computer with highest productivity; 1a – SP (single processor), scalar; 1b –
multiprocessor, scalar; 2 - personal computer; 2a - SP (single processor), scalar; 3 –
super computer; 4 - multiprocessor special computer; 5 – cryogenic SP (single
processor), scalar.

Fig. 3.1 Linear growth of computer processing power calculated from 1950 to 2002

can all be regarded as subsystems of TS. Each subsystem may be a single element or
group of mutually interrelated elements. Engine as subsystem has outer cover,
piston, cylinder and fuel injectors as component parts. Each part is made of one or
more substances. Outer cover can be aluminum coated with paint on outside and
titanium from inside. Aluminum and titanium are substances.

The process of TS expansion, i.e. first half of the wave of evolution, most
frequently starts from perceived limitation of substance. Right on level of substance, much stronger action of factors preventing increase of MUF is displayed.
Many big inventions and improvements through creation of new subsystems to
fulfill additional useful functions and increase of the existing function appeared for
lack of required properties of substances (materials) of TS, or for inability to use
hidden (not evident) resources (properties, effects) of the substance. The process of
TS convolution, i.e. the second half of the wave of evolution consists in a victory
of substance over subsystems. It is worth examining this paragraph in depth and
elaborating it step-wise.

We now define several stages or transitive moments in the full wave of TS
evolution: expansion followed by convolution:

(a) Attempts to improve (allocate) required property of the substance
(b) Division of the homogeneous substance into functional sites
(c) Specialization of sites according to their functions: transition to the hetero-
    geneous substance
(d) Compound substance made of specialized substances with high value of useful
    function
(e) Expansion of the compound substances to the subsystems
(f) Convolution of the compound substance or subsystem into an ideal substance

Above steps deserve some details with supporting examples.

TS is faced with a challenge to increase MUF. Apparently, the simplest solution is to increase MDE of a key substance of which TS is made of. Either or all of these factors, Mass, Dimensions (say thickness), Energy consumed (power supplied) can be increased. Remember, we should not be worried about rise of MDE at this stage. Because we are aware that TS has to expand initially to evolve. However, attempt to escalate MDE usually faces a contradiction—other properties or parts of TS start worsening.

Innovator now turns his head from 'headless MDE increase' to improvisation of a required property of the key substance. 'Depuration' begins, allocation of the required property, quenching of harmful incidental properties. Thus, many variants appear in form of modifications and models for different systems, objects, and work conditions. For example, over 3000 steel grades are produced in world today. Such super-specialization of the substance, nearly a new steel grade for every new TS, is forced.

If there is no opportunity to create a material covering the whole range of required properties, then one has to think microscopically. Anisotropic substances like crystals, wood, etc. are suggested for use with their 'right' orientation. In one of inventions related to 'forming roll for cluster mill', a remarkably high value of MUF that chiefly consisted in quality of rolled metal, was obtained by carefully employing mono-crystal of leucosapphire with its crystallographic axis oriented along the pivot pin of roll. In many industrial applications, a precious stone of first class—a sort of corundum, is used with most profitable orientation of crystal lattice. In processing hard metals, nature's hardest available substance viz. diamond is used with most 'cutting' orientation. How can we go beyond that? How do we produce harder substance than rightly positioned diamond? Does it mean end of rise of MUF in cutting of materials?

No, process of evolution now proceeds with division of mono-substance to sites, layers, parts resulting in transition to a compound substance. The reason is clear: at recurrent effort to increase MUF one soon finds out that whole substance need not have the property, on which the increase of MUF is dependant. Only a part of substance ought to possess this property. Engineers call this pampered part as 'working site'. It's easier to strengthen a desired property in working site than in the whole substance. Invention of smokeless blasting powder and introduction of rifled guns in the middle of the nineteenth century made a breakthrough in the artillery engineering. This was in response to a real opportunity to rapidly increase range of guns. But increase of power of the charge led constructors to the deadlock. Even change of copper and iron for steel did not bring the required result: steel pipes would stand pressure of maximum 2,000 atm. Moreover, thickness of walls had little influence on steadiness of stock. Only researches of Frenchman G. Lamé brought clarity: he showed that in the pipe influenced by equal pressure inside, layers of metal are given unequal pressure: interior layers bear basic tension, and exterior ones almost does not work. Consequently, there is no use in creating tools with very thick walls, unless one makes exterior layers work. The problem was wittily solved in 1861 by a Russian engineer A.V. Gadolin. He suggested strengthening the stock with

Fig. 3.2 Inspiration from nature

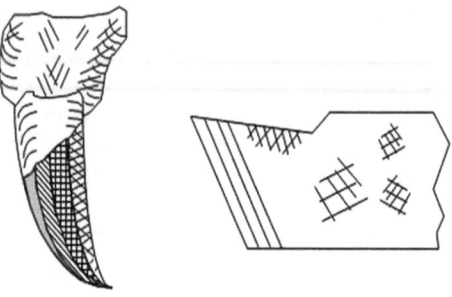

Principal of claws and fangs self-sharpening lies in the base of foliated cutters by M.Ignatiev.

rings—in hot condition they put cylinders on the stock, which after natural cooling pressed interior layers.

Recently investigation of a Russian sword belonging to tenth century but found in 1900 has been carried out. It was made of heterogeneous metal: cutting edges had a layer of advanced hardness and central part of blade was iron with low hardness. Micro-structure of the sites was also different, because these two materials were combined by blacksmith's welding.

Famous self-sharpening knives by A. N. Ignatiev invented way back in 1926, consist of several layers. A biologist by education, he wondered 'why fangs and claws of the animals are always sharp?' as they should grow blunt when they fray. Nevertheless, they remain not only sharp, but does not even change the angle of chock thinning. As it turned out, cause of it is different hardness of the interior and exterior sides of the chock. Less hard interior side of the fang frayed faster than the exterior one. For this, a sharp chock appears with constant effective angle of thinning Fig. 3.2.

The fact of transition from mono-substance to foliated one turns out to be useful. If definite properties are applied to every layer, one can get significant gain in MUF. Harsh winter jackets composed of multi-foliated panes have been made out. They are ten times thinner and lighter than homogeneous ones (even MDE has decreased here) and at same time, they offer better thermal isolation (MUF has obviously increased). Same method is used to quench wave processes, acoustic, optical, radio physical-elastic, etc.

In Donetsk Polytechnic Institute, researchers thought like this about machine tools: pivot of the arbor should have advanced durability and arbor should work properly at alternate load. Consequently, there should be more chrome and molybdenum in the pivot and nickel in the middle part of the arbor. Ideally, every detail should have some kind of mosaic construction. In any site of it, chemical composition and properties should be in accordance to the character of loads. Researchers succeeded in manufacturing metal tools and products with physical-mechanic properties varying throughout volume continuously (gradually) or discreetly (at once completely). These properties as functions of volume were set according to condition of operation of these parts.

After the division of the substance into the functional sites, a new process called 'specialization' begins. Each site fulfills only one function. No site is redundant. At specialization, it is easier to provide increase of useful function of every site and whole technical object. Italian firm 'Pirelli' worked out car tire with asymmetric tread pattern, providing equally good cohesion when driving on the snow or ice and on plain dry road. Such a tire is as if sealed of two different halves. A half set on the side of the car has a protector to drive on snow and ice and is made of rubber, having more silicon to provide better cohesion with the surface of the road. Exterior half of the tire has a protector to drive on the plain dry road and its rubber has more gas black, which makes better conditions to drive at high speed. Regardless of asymmetry of construction and varied composition of rubber, such tires get frayed equally. The firm guarantees a very long run before attrition.

Car headlamps are set to light road in front of the car. Taking safety into account it would be useful to have one more headlamp, which would spread light a bit up and aside, lighting the sign boards, standing on the way side. In UK, both functions were conjoint in one headlamp. A shoulder in the form of the prism was put on interior side of headlamp glass. Prism is such that at switching to anti-dazzle mode a part of beam of light of the headlamp declines aside and up, lighting side boards at a distance of 25 m from the car. A more serious problem of night driving is dazzle felt by driver due to light from lamps of oppositely traveling vehicles. Hundreds of patents were granted internationally on methods of prevention of dazzle but not universally acceptable and cheap technical solution came up. Some innovations offered 'differential' glasses or filters as windscreen or driving spectacles. But all these lowered visibility. Some other innovations used photodiodes in controlling brightness of one's lamp glow at oncoming 'rival' light current. Some innovations involved quenching the reflector with chokes, but these required a complete reconstruction of headlamps. Moreover, they were complex and not quite reliable. Some patents suggested polarizing glasses and filters: their usage meant 4-fold magnification of light power. Moreover, such glasses were quite expensive and in deficit. Some highways have oblique reflectors at headlamps' height installed in a row on the highway divider. Something like titled signboards on lamp poles. Only way to fight dazzle seemed switching to anti-dazzle. Anti-dazzle mode, popularly called 'dipper' is a universal mechanism installed in all cars around the world; when two cars approach each other, both drivers dip their beams and convey mutual respect.

Eventually, this problem was solved by Patent 520 487, which suggests a headlamp, which somehow bends light flow and it does not make the oncoming driver dazzle Fig. 3.3.

A French patent on application of optics is shown in Fig. 3.4. This is a method of defining esthetic properties of liquids contained in plastic and glass reservoirs. In the walls of the flask or bottle different optical elements like lens, prisms, etc. are formed.

In USA, a 'holographic window' was invented. Definite holographic structure was applied to the glass, with the help of which parts of the room that are usually dark, were lit. Such glass cover could direct sun light up to the ceiling instead of

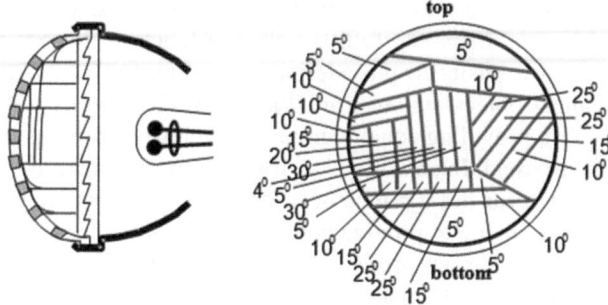

Model of headlight with 'prismatic echelon", Patent 520 487

Fig. 3.3 Technical solution to dazzle during night driving

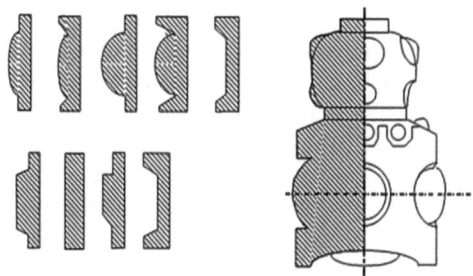

Method of application of transparent liquid due to light focusing through
the reservoir (Patent France 2 595 659) is shown.

Fig. 3.4 A French esthetic patent

the floor, light dark corners. Filtration of IR-rays keeps freshness and coolness in
the rooms. Sun light can even be transferred to the room without windows through
the air channel with reflecting walls and then made to disperse through the hole in
the ceiling.

Specialization of sites according to fulfilled functions leads to division of
heterogeneous substance into constituents, to replacement of separate parts by
substances with high value of useful function. For example, one of the modern
compositions of the kettle has an inbuilt three layer body: copper base for high
thermo conductivity, a thin inner layer of Teflon so that scale does not stick to it
and outer electrochemical layer that glows and provides safe covering. In Japan, a
new file of cheap unordered steel was worked out, cut of which was covered with
super hard ceramics (vanadium carbide). An outer layer to protect the file from
corrosion, to enable it to process hard alloys and to increase its life duration by 5–6
times was coated. In France, lead-acid accumulators were produced. Their weight
was 4 times less than usual ones, as they contained only a functional layer of lead
(lead is heavy, so was trimmed) in them that was applied to glass and carbon
fibers.

In cities, glass surfaces get dirty so often that even constant washing cannot keep them clean for a long time. A composition 'Isolver' was invented in France to help solve this. After its application on clean and dry glass surface, it prevented adhesion of rain water, sticking of polluting particles, formation of frost, etc. it. Composition was chemically neutral and a liter of the composition was enough for 100–120 square meters of the surface.

Substances fulfilling the required function on their own, i.e. working on their own energy or that available in the system are not always available or producible. In this case, a servicing subsystem joins the substance. Both drivers and pedestrians know it is not that easy to distinguish traffic light signal on a sunny day. Reflecting from color glasses sunlight gives wrong signal. For this reason, patent on traffic lights with black curtains appeared: when the lamp (for example red one) is turned off, its glass is covered with automatic curtain. According to Great Britain Patent 1 454 386, glass of the lamp is covered with film of liquid crystals with electrodes on the sides; when the lamp is 'switched off', liquid crystals does not let the light pass and look like lustreless black surface; at turning the lamp on electric field produced by flowing current reorients the crystal molecules and the curtain becomes transparent.

Sooner or later, subsystems or compound substances should again convolute into substance. Such substance that experienced a circle of expansion-convolution and got new quality, providing high value of MUF in the concrete TS can be called an ideal substance of the first range (IS1).

It is known that ultraviolet light oppress plants. Especially, greenhouse plants are sensible to ultraviolet light. Bearing this in mind, specialists all over the world cover roofs of the greenhouses with a light filter film also. Ultraviolet is absorbed and transformed into heat. It is also found that light with the length of the wave lying in red–orange area has good effect on all plants. They transform it better into chemical energy in the process of vital activity. But it is just impossible to cover greenhouses with one more film: optical transmission will immensely drop. The attempts to apply two opposite properties to the glass, viz. not to pass UV and keep IR light, failed. The problem was successfully solved by M.S. Kurnakov. A film transforming UV light into IR light was created. Herein, harmful factor is taken away and the useful one is introduced simultaneously. This transformer film used luminophor on the basis of europium—micro doses of luminophor are mixed into polymer. 'Polysvetan', as film was called, gave an unexpectedly high increase of harvest: tomatoes and cucumbers—50 %, salad—20 %, water melon—60 %! Needless to mention, Polysvetan is an IS1.

Tracing Ideality in Expansion-Convolution Waveform

Applying I(S) to expansion-convolution waveform, I(S) may or may not increase during expansion, but it definitely increases during convolution. This is an important key to locating current position of a TS on real-time axis. But since every expansion is followed by convolution, it can be stated that Ideality must increase for system to evolve. So here comes the Law of increase of degree of Ideality. It is 8th and last of our 'Laws of Technical Systems' Evolution. Restating this important law:

Evolution of All the Systems Moves in the Direction of Increasing Their Degree of Ideality

An ideal TS, also called an IFR, is characterized by an infinite or near infinite value of Ideality. In words, an ideal TS is a system whose mass, dimensions, and energy consumed (MDE) tend to zero, but its capacity to fulfil work does not decrease. In the limit, an ideal system does not exist, but its function is preserved and fulfilled. In reality, non-existence is impossible. A near-ideal system is one with very nominal value of MDE but a substantial (required) value of MUF. Two examples given here:-

Example 1: Gas leak detected cured by microphone

Salient comment: Solution close to IFR achieved through replacement of expensive sophisticated measurement TS by readily available, cheaply priced TS. MUF is maintained constant.

A gas pipe was suddenly found leaking in a chemical plant. Gas was nearly transparent so visual detection was unreliable. But was dangerous by being inflammable and toxic, so urgent action was mandatory. The engineer on duty rushed to emergency control room to fetch a gas detector. He was unable to find it. What did he do? He looked for an ordinary microphone in store room and luckily found it. It was run over doubted part of pipeline. Location of 'hiss' identified leakage. A two Euro microphone saved the plant. A significant suggestion emerges from this construction and usage of this near ideal TS.

Identify all available fields, substances, and their interactions. Here, audible sound waves, concentration gradient field (causing diffusion and undesired spread), mass transfer (leakage into environment) are available resources. The first one was picked.

Example 2: Fuel tank of spaceship:

Salient feature: IFR may seem physically impossible or unattainable, but it must be held in mind while idealizing TS toward solution.

In space, in condition of weightlessness, fuel in tanks of space vehicles breaks down to drops. The big and small balls freely float inside space of tanks. They may not be present near intake aperture; fuel cannot be delivered to the chamber of combustion. What was the solution adopted?

Following solutions were initially offered:

1. Before turning on the basic engine, an additional micro-motor (for example, using gas) is turned on for some seconds. This micro-motor push the fuel; it nestles to the intake aperture of tank. Solution is unacceptable because it complicates control system and reduces reliability of entire space vehicle.
2. Tank has two compartments separated by an elastic partition such as a well-defined membrane. On the one hand is fuel, while on the other hand is gas.

During refueling at the cosmodrome, gas is compressed. In space, under the action of gas, membrane pushes fuel to the aperture till tank is fully emptied. This offer was also rejected. Fuel during refuelling should be delivered under pressure to compress gas behind the membrane. Since pressure is higher, fuel contains more dissolved gas. In an orbit, when fuel enters chamber of combustion, dissolved gas allocates and form bubbles. It can result in damage of space vehicle. Preliminary degassing of fuel before refuelling is not recommended because degassing causes easy fractions of fuel to disappear. So no option could make this solution acceptable.

IFR: fuel itself moves to an intake aperture and always positions near it. Pay attention: reasoning whether it is possible or impossible that fuel itself places near the aperture, and if yes, how it will be carried out has no influence on formulating IFR.

Adopted TS: Figure 3.5 has illustration and explanation.

All systems, irrespective of domain they belong to, tend to achieve IFR. So much so, we have an alternative, more explanatory name for TS evolution. It is idealization. We have learnt earlier that evolution of TS consists of alternating phases of expansion and convolution. This section is devoted to tracing movement of factor of Ideality, I(S) as TS travels its interesting evolutionary path. We achieve this by choosing two case studies: Refractrometer and Mirror for a powerful laser.

Case Study 1: TS is Refractrometer

Refractometer is a device to measure value of light refraction in a transparent or translucent medium (substance) and hence deduce the refractive index of that medium. Why is this measurement so important? Any change in value of index corresponds to changes in several physical–chemical properties of the substance. Hence, accurate measurement of this index is important in controlled processes.

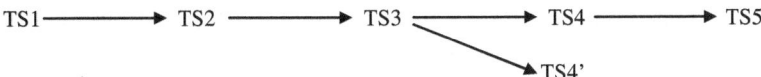

TS1: represents refractrometer in its simplest form Fig. 3.6. Mechanism of measurement used is unchanged since ancient times: one makes a prism of a substance if substance is solid or pours it into the glass prism if substance is liquid and passes a light ray through it. The ray declines. This declination is recorded and a concrete optical property called refractive index is defined. Refractometers can measure tightness and concentration of solutions, pulps, and suspensions; hence their use to control processes in tinned, sugar, alcohol, and other industries.

Tank of aerospace vehicle (USSR Patent 731886)

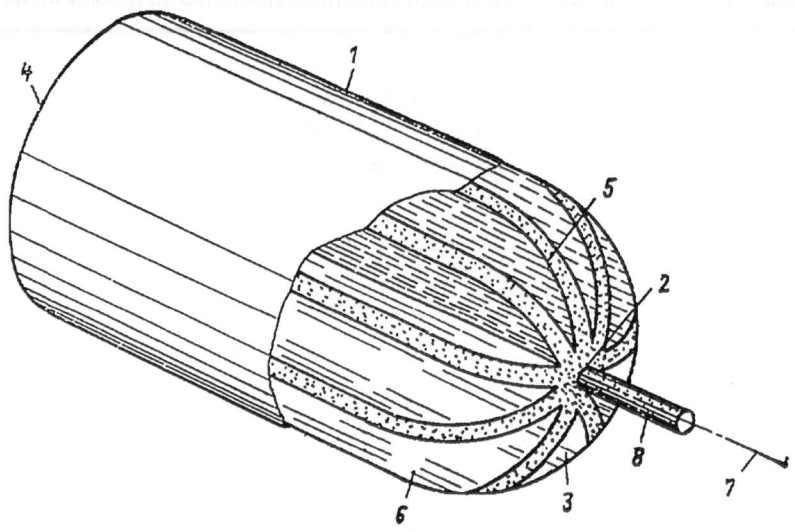

A tank has cylindrical case 1 with drain hole 2 made in the bottom 3 (section of the bottom decreases gradually). Opposite flat bottom 4. Inside the case there is a block 4 made of porous material (as felted cloth on nickel carrier) in which longitudinal channels 6 made. Porous material made as hollow chords. Diameter of pores decrease along the direction to the drain hole. Channels 6 have cross section as round sectors directed by angles to the longitudinal axis 7 of the tank. A wick 8 made of porous material is entered into drain hole.
Fluid discharge out of tank in the weightlessness condition occurs by capillary forces acting along the direction of constriction of tubular block 5 pores. Fuel moves to the drain hole 2 and through the wick 8 delivers to the engine

Fig. 3.5 A Soviet patent in space technology

In aviation, refractrometers can sensitively determine and control contamination of water in fuel while it is stored in tanks.

TS2: TS1 appears as highly 'expanded state'.[1] So system begins its evolution wave with convolution. Prism is absorbed by light conductor. Or otherwise: light conductor, which earlier only conducted light, now fulfils prism function. Whatever view is selected, two components blend into one. Light conductor of approximately U-shape is immersed into liquid. Light is sent from light source at one end. It is received by a photodetector at opposite end. Percentage of light energy which is successfully transferred depends on refractive index of medium.

---

[1] *Several versions of TS exist before TS1—these are not discussed here. Our start point is TS1.*

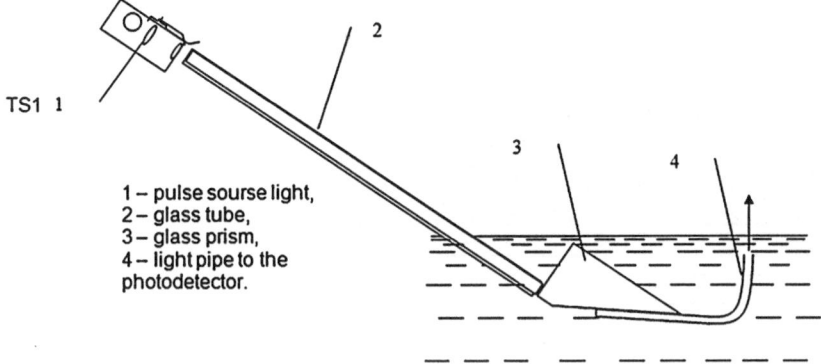

TS1 1

1 – pulse sourse light,
2 – glass tube,
3 – glass prism,
4 – light pipe to the
photodetector.

Refractometer is immersed into the solution to measure density. Light current changes
through the prism depending on liquid density.

TS2 2) Czech patent 124740

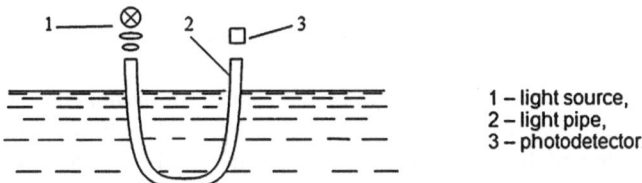

1 – light source,
2 – light pipe,
3 – photodetector.

Fig. 3.6 Refractrometer's evolution in progress

Where is remaining energy lost? Part of it is lost to medium, part of it is absorbed. Disadvantage: change of liquid level causes light current to change Fig. 3.5 again.

TS3: Light conductor is enclosed in safe covering with only the lower (measuring) part uncovered. TS2 expands to TS3. Still, few disadvantages remain: exactness of measurement depends on diameter of optical fiber and in the areas of light conductor bending, there are substantial light losses.

TS3 evolves either to TS4 or TS4'. While TS4 can be considered as next step to TS3, TS4' is a branch line end product for specialized applications like airplanes.

TS4': Figure 3.7 is a severe case of expansion again but should not be so annoying if huge gain of MUF in terms of increased sensitivity is accounted. Refractive index of fuel in tank (1) of airplane is to be measured. Light conductor is made in the form of plate. Light travels from source at one end to receiver at other end via a highly efficient (up to 100 %) mechanism of total interior reflection. Electro-optical material changes the value of refraction under the influence of electric field. Receiver transforms light into electric current. Block of control directs the current to electrodes of liquid crystal and changes the refraction in it—till it is equal with that in liquid. Exactness—on hundredth and thousandth part of coefficient of refraction can be achieved. The disadvantage: instability of device

4) Patent 840711: TS4'

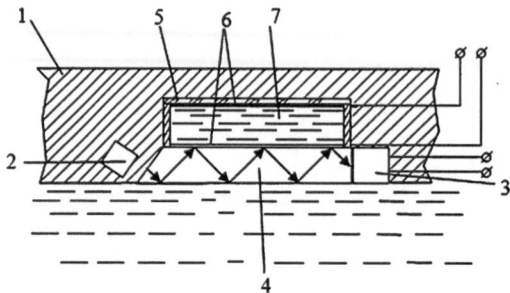

1- wall of the fuel tank of the plane, 2- source of light, 3- light receiver, 4- parallel-sided plate (light conductor), 5- glass lid, 6- transparent electrodes dusted on the glass, 7- electrooptical material (liquid crystal).

Fig. 3.7 Specialized version of refractrometer to measure fuel purity in airplanes

characteristics when subjected to exterior influences like temperature variations and electro-magnetic irradiation.

TS4: Poly-system with biased characteristics: case of partial convolution: Patent 994965. In Fig. 3.8 there is only one source of light and multiple light conductors. Refractive index of liquid (6) has to be measured. Main light conductor has a linearly descending refractive index along its length. Series of output prisms with light conductors are installed along main light conductor at intervals of 5 mm. Light spreads from the source along the plain light conductor in the form of successive complete interior reflections. Through the output prisms and light conductors, light comes to the indicator scale. Refractive index can be higher, equal, and lower at different sites of the plain light conductor. At site where the value of refractive index of light conductor is lower than that of liquid, light moves into liquid and indicator scale is dark there. Location of border of light and shade on the scale defines refractive index of liquid. Exactness of measurement is one degree higher than best refractometers. Added advantage: protection against electromagnetic radiation as well as thermal compensation is not required.

TS5: Poly-system with more idealization: high level of convolution: Patent 1225355 Fig. 3.9.

Operating principle: for every liquid, there exists radius of bending of light conductor at which, condition of complete interior reflection in the light conductor is broken. Ends of the light conductors with radius of bending lower than this critical radius would not shine while ends of the light conductor of a bigger radius will shine. Ends of the light conductors make an indicator scale. In this highly convoluted poly-system with biased characteristics, we have one light source is left and a beam of differed light conductors.

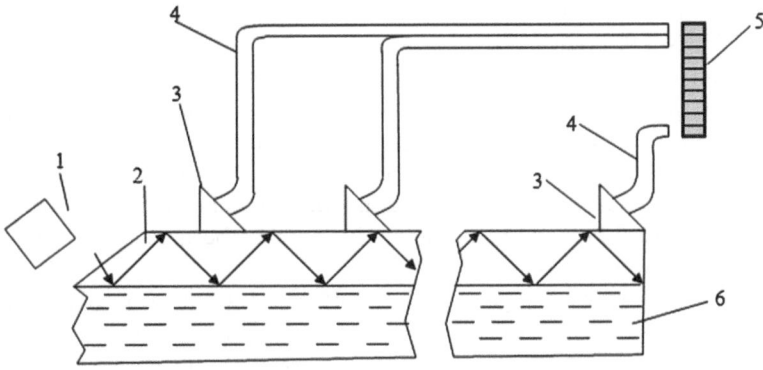

1 - source light,
2 – flat light pipe,
3 – prism for output of radiation,
4 - light pipe,
5 – indicated scale (flanks of light pipe),
6 – controlled liquid.

Fig. 3.8 More progress

Fig. 3.9 Highly idealized machine

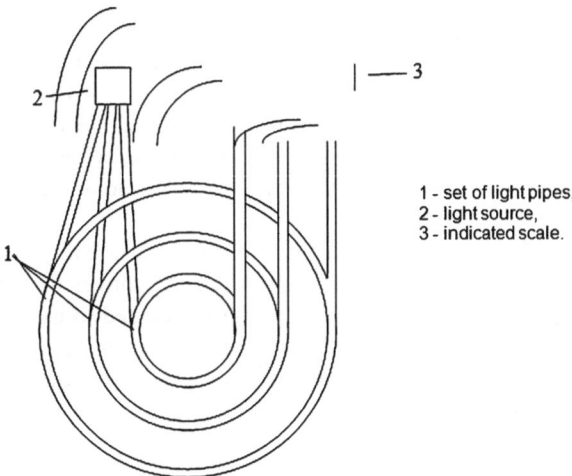

1 - set of light pipes,
2 - light source,
3 - indicated scale.

Case Study 2: TS is Mirror for Powerful Laser

Lack of knowledge about such regularities costs society a lot, as years pass before world of inventions face progressive technical solutions connected with usage of CPM and effects accompanying these substances. Quite demonstrative, from this point of view, is the story of working out new types of mirrors for powerful lasers in the Institute of general physics Russian Academy of Science ("Science and Life", 1985, NO 9, pp 50–53).

Resonator, consisting minimally of two optical mirrors, has been an integral part of most lasers. Through one of them, e.g. half-transparent one, radiation is lead out. The first generator mirrors were traditional—quartz discs with silver covering. But in recent years, power of lasers has grown by hundred, thousand, and perhaps million times. A problem of creating mirrors, being able to work under the influence of powerful radiation appeared. This problem became one of central ones in improving powerful lasers.

A mirror even with very good optical surface does not fully reflect radiation falling on it, about 1 % of it is absorbed and turned into heat. In powerful lasers, this 1 % is enough for thermo tensions to appear in the mirror. They distort geometric form of the reflecting surface disabling fine focussing (and hence desired concentration) of rays. In fact, thermo deformations lead to break of phase; laser stops being a laser.

What is the limit for these troublesome thermo deformations? It should not exceed 5–10 % of the length of the wave of laser radiation. For CO2 laser with length of the wave 10.6 μm falling in infrared band, distortions should not exceed 1 μm. If one takes such a mirror in hands, in few seconds deformations of optical surface will exceed the permissible magnitude because of the unequal warmth provided by hand. But this is temporary 'spoiling'—these are elastically reversible deformations. Beside elastic deformations plastic deformation can take place at bigger powers, and then the mirror area would be destructed permanently.

Translating problem as power challenge, it was required to produce mirrors that withstood prolonged tensions up to several kilowatts to 1 cm^2 of their surface. This power can be compared with that radiated by the sun from its surface. It means that if we put our mirror on the sun, its form should not deform for more than a micron. Indeed a challenge!

Physicists considered it like this: quartz badly conducts heat, thus it should be changed for metal. For fully reflecting mirror it is okay. But for half-transparent mirror, they got stuck. They finally decided to use metal for this purpose also, but with a modification. A hole in the center of the disk let some radiation pass through. Resonator appeared like chamber with aperture at one end. Figure 3.10 traces entire evolution of this technical system.

(a) resonator assembly unit, radiation is removed through the hole in the mirror,
(b) mirror with channel structure, current of heat carrier cools ribbed wall,
(c) mirror is cooled by heat carrier, coming through the pored material,
(d) water is pumped to reflecting surface by canals, gets boiled, mixture of liquid and vapor is lead to transverse current of heat-carrying, is cooled and is gotten out of mirror.

Metal disks removed heat well but had disadvantages: high coefficient of thermal expansion altered their size and form at change of optical loading; low firmness so it is difficult to polish.

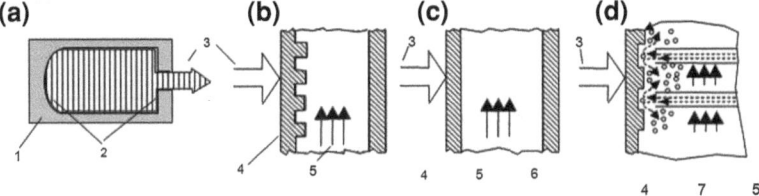

Fig. 3.10 Elaboration of a mirror for a powerful laser

Interlude: change of quartz for metal as technical solution seemed wild, strange; it just shocked opticians-specialists in producing mirrors. Little did they know about queerer surprises ahead.

Searching for a better composition began: metals, alloys. Almost all alloys available for mass usage were tried. As a result of such searching, increase of the limit of optical working capacity was managed by 10 times. Fight did not end here.

Requirement of power of light went up. Increase of MUF. With it grew thermo loadings. Metal thermo conductivity could not provide dissipation of this powerful heat current. How was this problem solved?

Cooling is required, a forced distraction of heat by some moving liquid. Greater is the difference of temperature between heated body (mirror) and cooling body (liquid), faster is shed of heat. Calculations showed that the problem would have been solved if magnitude of this difference was more than 1,000 °C. It means that mirror should have a temperature 1,000 °C above liquid. But such temperature is impossible for metal mirror as one cannot provide good quality of optical surface at such temperature. Contradiction: one requires high temperature for good heat distraction and for the stability of geometric form and other optical characteristics of the mirror one requires low temperatures.

Site specialization started with attention focussed on back of mirror—it was this part of mirror which exchanged energy with running liquid. Smooth surface on the back of the mirror does not provide the required intensity of heat distraction. In order to enlarge the surface of heat transmission, they undercut trenches along which water was led. To fasten heat transmission, walls of canals were thinned and velocity of water increased. This reached a limit too. Walls of canals trembled and deformed under water pulsation. This contradiction was resolved by making transition to capillary porous material, popularly known CPM. Advantages of CPM: vast surface of heat transfer, good intermixing of cooling liquid which moves in the capillaries, high mechanical tightness of matrix-skeleton safely carries the mirror surface and saves its geometry. CPM is applied with a covering and polished to turn it in a mirror. Thickness of the covering is 100–500 micrometres, not more, otherwise it would retain heat. Possible way of application is via chemical transportation reaction from gas phase, i.e. collecting at atomic level. And it means that the surface grown would primarily be smooth—humps and valleys not more than 0.1 μm. Post-processing viz. polishing, roughness remains only a thousandth part of micron.

Fig. 3.11 Designed for Mars

But again man is discontented. Laser becomes powerful, same old increase of MUF. With it, temperature rises. So speed of heat carrier motion is increased. How? Liquid molecules now instead of 'floating' must 'fly'. How to do this? Contradiction: for good heat removal, the agent should be presented in liquid state (high thermal capacity) and for quick exchange of heat (inflow-outflow at high speed)—by gas. Solution is rendered by use of 'resolution of physical contradiction by separation under condition'. Phase transition. At heat removal, it should be liquid, and at elimination—gas (vapor). Liquid should get boiled into vapor and latter should impetuously get away from the heating zone. To improve boiling, it was done under pressure of air. Good part is that air molecules does not interfere with vapor particles movement. Heat pipe! Yes. As liquid they used melted metal, which took quite big portions of heat with it at vaporization. Vapor speed reached sound speed but this was the last border.

Till now, on such mirrors great intensity of thermal dissipation was achieved, up to several tens of kilowatts on 1 cm^2, actually it reached 100 kw/cm^2. And then? Increase of MUF is a non-stopping process. How can one withdraw power of 1,000 kw/cm^2? Or 10,000 kw/cm^2?

At such power, the thickness of the wall should be vanishingly small—1 μm or even 0.001 μm, i.e. there should not be any wall at all. And the mirror itself, of whatever substance it is made of on Earth, will disappear into gas or plasma at 10,000 C. So there is no mirror, but the function should be fulfilled. We are reminded of treatment of IFR earlier in book. The field (laser ray, electromagnetic radiation) should produce a mirror for itself. This mirror's surface made of liquid or gas has constantly renovated surface. Renovation is done by laser itself. Another physical limit, after sound speed, is more fundamental, viz. speed of light. Let heat be withdrawn at such speed by infra red radiation? Laser power can increase to several times. Story of TS development continues.

Problem 6: Biologists invented a new type of compact hydroponic device for supplying of fresh greens to cosmonauts during long expedition on Mars. Testing of device was planned on orbital stations in automatic mode without human's participation. For long time, engineers could not solve the problem of cooling the solution circulating in device (Fig. 3.11).

Plants produce warmth during their growth process. This heat has to be shed. It was calculated that an area of 1 m^2 was ample for an air-based heat exchanger. But periodically, during definite short-time periods, these plants produce heat 5 times more than usually. To install 5 such exchangers in parallel is impossible. The reason is not just weight increase. If solution is made to traverse 5 pipes, speed of

Fig. 3.12 Idealized radiation
measurer

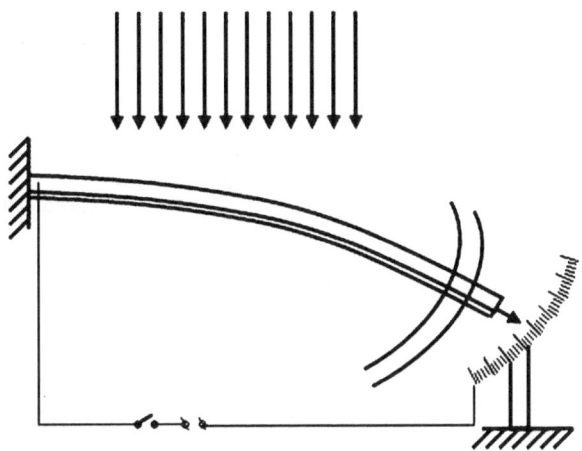

movement decreases by 5 times compared to loss of speed in single pipe. Also solid components settle on walls of tubes. Hence, this multiplication solution was inadmissible. Enlarging length of single exchanger by 5 times is also highly inefficient; hydraulic resistance, the expense of energy, weight, etc. all increase impermissibly. Requirement is straightforward: in these heat-bursting periods of time, area of exchanger should enlarge by 5 times. What was the final technical solution?

Heat exchanger was supplied with petals made of shape memory alloy, titan + nickel, pressed to it. At rise of temperature, petals are unbent, increasing the area of cooling by 5 times. Heat exchanger was 1x during normal operation and 5x during those short heat bursts.

Few examples of perfected or near perfected technical systems are offered; idealization in full swing is apparent.

Example 1: Radiation measurer: Patent 1026550 Fig. 3.12. Dosimeter in the form of plate of mono-crystal with applied film. Mono-crystal and film have different values of radiation induced coefficient of elasticity. Influenced by radiation, bi-system, viz. mono-crystal + film, bends. An arrow on one end indicates magnitude of radiation on scale. Physical effect: influenced by radiation, tightness of materials, and coefficient of elasticity change linearly in the angular range of bending involved. To regenerate properties, element is heated for 5 min at temperature of 600–700 °C by current from the source. After annealing, the plate is straightened. A pair of materials used: BeO-Al2O3, SiO2-Si, etc. Here technical system, radiation measurer, is totally absorbed by an ideal substance.

Example 2: Fuel tanks in wings: In the construction of the plane ANT-25, Tupolev managed to get gain without losses. Two big riveted fuel tanks, for which there was no place due to their hugeness and weight, were put inside both wings. Each tank was extendedly shaped, i.e. placed along the whole wing. Wings during flight suffer from strong tensions, from appearing air dynamic powers. These powers are directed upwards. While force of tanks, their weight, is downwards.

Fig. 3.13 Tupolev in action

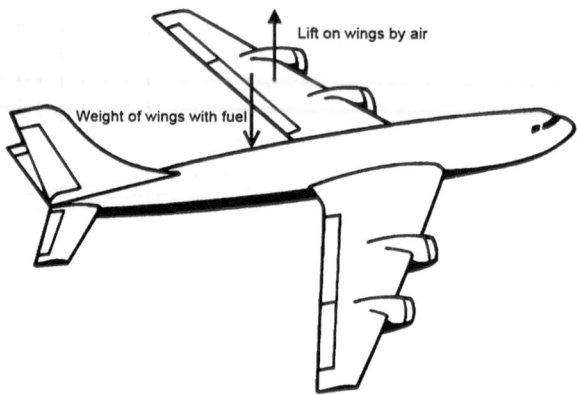

The two compensate one another. It turned out that due to unload of the wing, one could decrease weight of metal used in wing and tank. Total weight of the plane decreased, its speed increased. *This example show how superposition of subsystems leads to idealization. Useful properties are mutually added* (Fig. 3.13).

Problem 7: There is a pool with water. Area: 25 m squared, Depth: 10 m. Temperature of upper layer of water is +900 °C while that of lower is +100 °C. What is the particular way through which upper and lower layers are heated to these temperatures is not important. One has to provide equal temperature in the pool. One cannot use pumps, activators, etc. This is a nuclear power station, dangerous, a site by the close to reactor; one cannot uses any equipment as it requires repairing, service. One cannot also apply thermo-emf based mechanisms due to their low coefficient of efficiency. How to solve the problem? Remember, according to the laws of physics hot water rises up, cold water sinks down. And it is got to be done vice versa for solving this problem.

Solution: Witty convolution was found: hollow Ni + Ti balls are used with memory programmed on two conditions—sphere at 100 °C and pressed sphere at 900 °C. Spherical balls float while pressed balls sink in water (Fig. 3.14).

Problem 8: The most dangerous thing for sky-scrapers at fire is overheating of steel carcass. Carcass metal loses fastness, becomes plastic (in short, ruins) at reach of definite temperature. How to prevent this? Usually the carcass is made of hollow steel pipes and cut-outs like ones shown in Fig. 3.15.

Very conventional solution: One usually covers steel constructions outside with fire resistant material and face with steel or aluminum. This is an expensive and long process. Moreover, such 'sandwich' does not survive at prolonged local influence of fire.

An existing solution: This is based on an adopted architectural design in the US. Underlying principle is to neutralize surplus of the field (thermal) by a substance. The frame of the building is made of hollow steel structures filled up with water and connected with an expansion vessel on the roof. Salts of potassium, to retard corrosion of metal are added to water. Water freely circulates in the frame. During an unfortunate fire, if boiling water and steam builds up a high pressure, a safety

Fig. 3.14 Heat flow
'against' laws

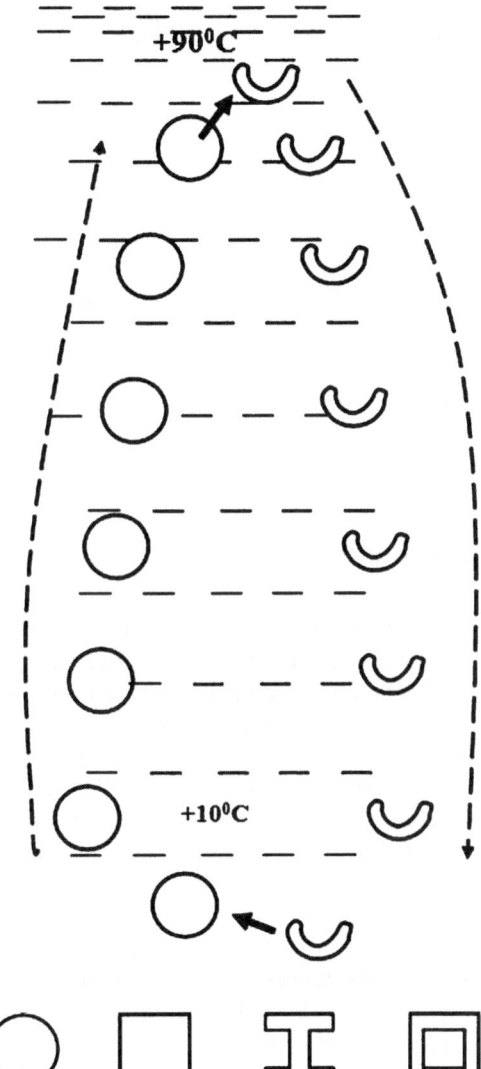

+90⁰C

+10⁰C

Fig. 3.15 Profiles of pipes in
skyscrapers

valve turns on to release them to the atmosphere. Weight reduction of columns and
lower cost of construction are achieved due to exception of protective layer.

More ideal innovative solution: The underlying principle is ancient, universal
one: ample water should be there where there is fire. Water itself should move
there. It should move faster, when and where strong heating and evaporation
occur. Hollow constructions have an internal covering of capillary porous material
(CPM). At evaporation, the water immediately moves out from all sides. Steam
freely moves on an axis of construction upwards. Service and operators are not

Fig. 3.16 Theoretical
idealization of systems

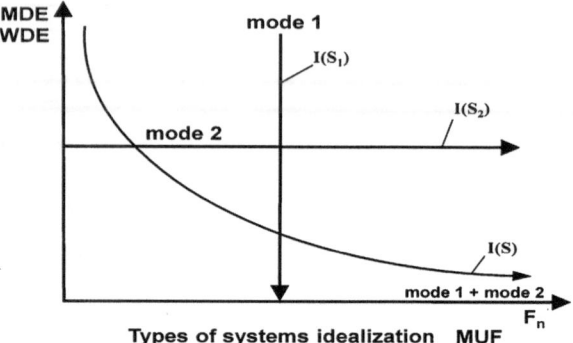

necessary. We can call this machine as a specially designed thermal pipe. Convolution of high degree has taken place.

Convolution: A Graphical Perspective

To fulfill a function, a material object is required, whether this or that. So for TS to vanish or contract, other systems (adjoining TS, super or subsystems) should fulfill this function instead of the disappeared (idealized) system. It means that a part of these systems is transformed so that they should fulfill even additional functions—those of the disappeared system. If 'alien' function taken to be fulfilled is similar to its own, then simple increase of MUF of that system occurs; if the functions does not coincide then an increase in number of functions of that system takes place. Note that the other system, adjoining, super or subsystem is being referred to in previous sentence.

To sum this—

Disappearance of systems (MDE) and increase of $F_n \sum MUF$—are two sides of the common process of idealization. These two sides *can* occur individually—leading to two distinct types of idealization (Fig. 3.16).

Mode 1-I(S1). Idealization of first type, when MDE, viz. mass, dimensions, and energy consumed tend to zero and $F_n \sum MUF$ remains unchangeable.

$$I(S1) = \lim_{MDE \to 0} \frac{F_n \sum MUF}{MDE} \left(F_n \sum MUF = Constant\right)$$

Mode 2-I(S2). Idealization of second type, when $F_n \sum MUF$ increases and MDE, viz. mass, dimensions, and energy consumed remains unchangeable.

$$I(S2) = \lim_{F_n \sum MUF \to \infty} \frac{F_n \sum MUF}{MDE} (MDE = Constant)$$

Fig. 3.17 Practical
idealization of systems

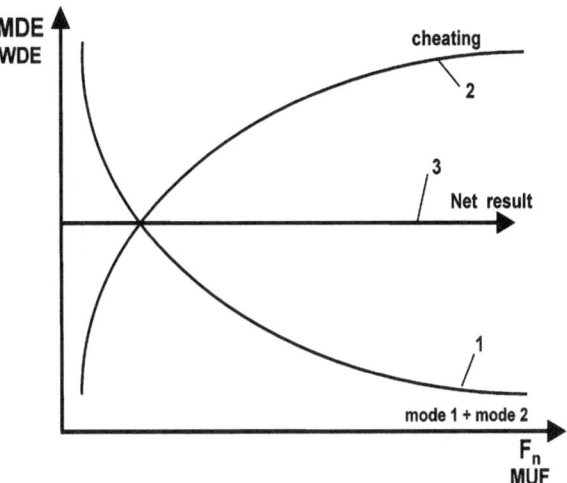

$F_n \sum MUF$ is either sum of MUFs or a function (derivative) of MUFs. If MUF is single, it can be written as MUF only.

Idealization *can* thus occur via mode 1 or mode 2. Can mode 1 I(S1) and mode 2 I(S2) occur simultaneously? Of course, yes. In this mixed yet best type of idealization, both processes viz. decrease of MDE and increase of MUF is affected. We call it simply I(S).

$$I(S) = \lim_{\substack{F_n \sum MUF \to \infty \\ MDE \to 0}} \frac{F_n \sum MUF}{MDE}$$

Thus Mode 1 I(S1) + Mode 2 I(S2) = Mixed Mode I(S)

It means that extreme case of idealization of engineering concludes in its decrease (ultimate analysis, its disappearance) when at the same time the amount of functions fulfilled by the system should increase; ideally, there should be no engineering and functions which mankind and society needs should be fulfilled.

But as they say, theory is far away from practice!

Idealization of real TS follows way, which is different from mentioned above. Real TS follow as in Fig. 3.17 shown.

Real TS takes on curve 1 which is identical to I(S) = mode 1 of idealization of hypothetical TS + mode 2 of idealization of hypothetical TS. But all is not glittery. Subsystems, peripheral systems, super systems start developing or increasing in numbers. Curve 2 represents this 'cheating'. To count everything honestly, this cheating must be accounted for. In other words, curve 1 showing MDE versus MUF must be added to curve 2—MDE versus MUF of surrounding/sub/super systems. Curve 3 shows this final effect.

$$\text{Curve (1)} + \text{Curve (2)} = \text{Curve (3)}$$

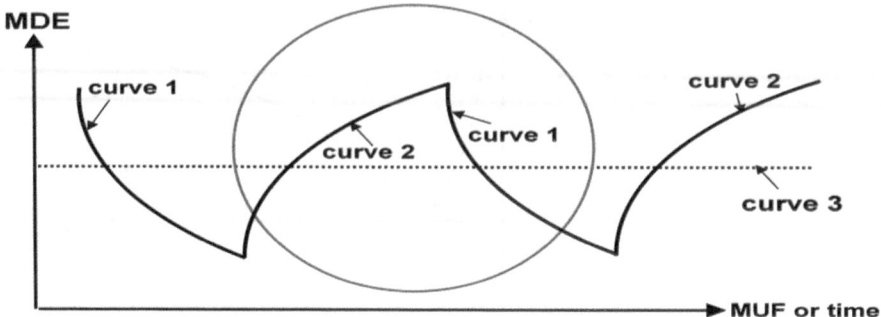

Fig. 3.18   Continuous waveform of expansion, contraction, expansion

Fig. 3.19   Singled-out
waveform

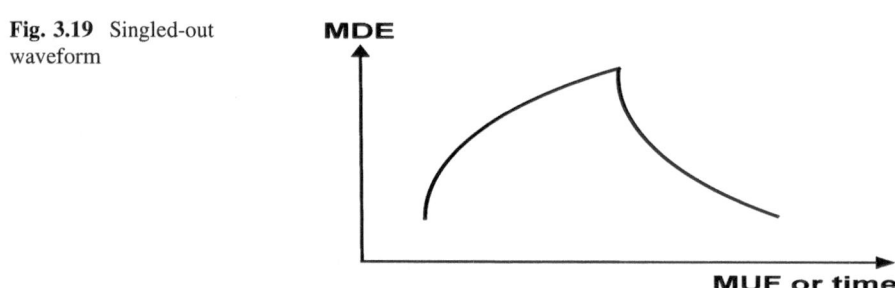

It is surprisingly straight. Note: it resembles mode 2, I(S2) of hypothetical system idealization.

Real idealization (net) → MUF increases at constant MDE

Why does this occur technically? Gain in MDE, technically, its numerical decrease, which is obtained in the process of idealization is spent immediately on creation/addition/development of additional systems—adjoining, super or subsystems. Aviation, water transport, military engineering, etc. typically exhibit prominent appearance of curve (2) after curve (1) moves through.

Process of idealization of real TS is outwardly similar to the second type of I(S2), when increase of MUF takes place at unchangeable values of MDE. Set of subsystems as whole follow curve 2 designated cheating. It is interesting to point that MDE of individual subsystems decrease, but these subsystems are doubled, trebled, new ones appear, etc. Fig. 3.21. A subsystem individually follows first type of idealization I(S1) while summation of all subsystems follows curve 2 (cheating).

Curves (1), (2), and (3) are all processes in time. Curve 1 and 2 do not occur simultaneously, rather they follow one another alternatively. So TS (with its subsystems, etc.) goes along curve 1, curve 2, curve, and so on. Curve 3 now becomes the time-averaged graphical depiction of system (Fig. 3.18).

Encircled portion is isolated and examined now Fig. 3.19.

Fig. 3.20 Smoothed singled-out waveform

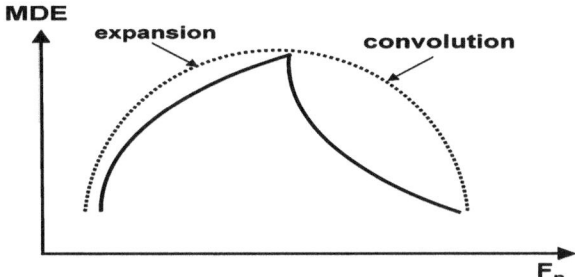

Fig. 3.21 Generic model of TS growth

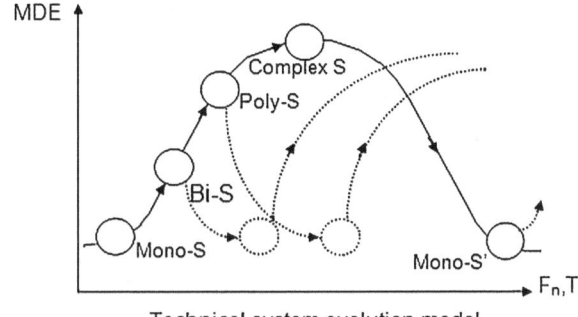

Technical system evolution model

An envelope curve over curve 2 and curve 1 is drawn, generating a dotted final curve (Figs. 3.20, 3.21). Notice, it has expansion followed by contraction or convolution. It is in latter part of curve, representing convolution that we are focusing upon in current book.

Chapter 4
Four Types of Convolution: Miniaturization Embedded in 2nd Type

As opposed to expansion, convolution, deeply and comprehensively captures the structure, organization, and system properties of TS. This period completely corresponds with the law of increasing ideality: TS decreases its MDE while simultaneously increasing its MUF. A TS after reaching a point of maximal expansion, chooses to contract (convolute) by any of four mentioned types

1. Displacement of a part of subsystem into super-system
2. Development of a subsystem belonging to TS
3. Convolution of TS into one of its subsystem
4. Convolution of TS/one of its subsystem into an ideal substance

In the development of real TS, processes of convolution can occur at any level of hierarchy. See Fig. 4.1. The layers in graphs from up to down are of : super-system, system (TS), subsystem, and substance. In different methods of convolution listed above, TS (or its part, say subsystem or substance) moves in different directions. See arrows in Fig. 3.21. The scheme resembles chaotic image of Brownian motion. Though there is an apparent chaotic state of TS development, final destination is same. Howsoever may TSA evolve, it has to reach final convoluted state TSB. So while this process is in tune with Law of Irregularity of System's Parts Evolution, Idealization is sure and follows Law of Increasing Ideality of technical systems.

All four ways lead to the same TSB characterized by small MDE and high MUF.

First way of convolution—displacement of a subsystem or its part (this subsystem belongs to TS) or out of TS border and its conversion into a specialized system which now becomes an integral part of super-system.

This mechanism is characterized by the following peculiarities:

(a) The amount of elements in TS decreases,
(b) MDE of given TS decreases,
(c) MUF of given TS increases due to action of two factors:

S. Kwatra and Y. Salamatov, *Trimming, Miniaturization and Ideality via Convolution Technique of TRIZ*, SpringerBriefs in Applied Sciences and Technology, DOI: 10.1007/978-81-322-0737-5_4, © The Author(s) 2013

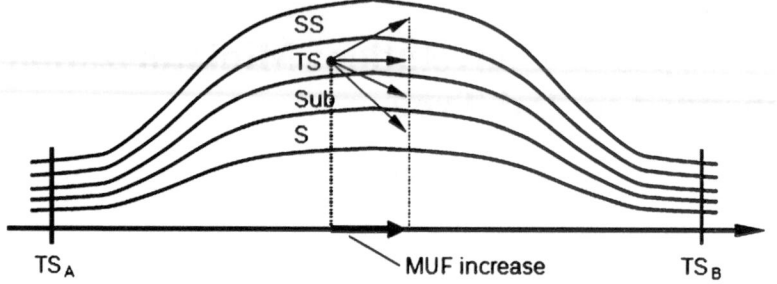

Possible ways of idealization of technical systems.

Fig. 4.1 Inverted Bell-shaped idealization of TS

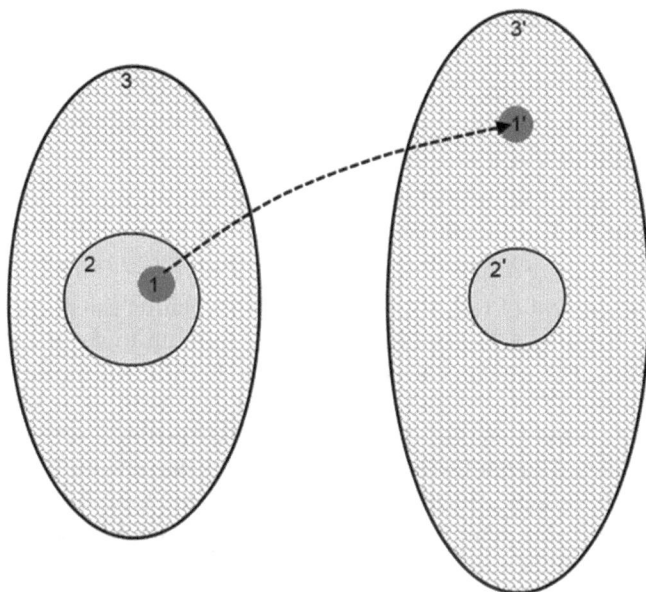

Fig. 4.2 Convolution of first type

(i) The system become lighter as it has to be universal, multi-purpose; the structure and organization become simpler, functioning becomes better.

(ii) The same function (action) of higher quality enters from super-system instead of the displaced subsystem. Remember that this former subsystem is now a specialized system in the super-system.:

There is a limit to which this method can convolute. The number of elements in the TS continues to decrease until working unit remains. Working unit cannot be eliminated, because its elimination implies termination of TS. The displaced subsystem after 'conversion' to a specialized system joins super-system to fulfill MUF Fig. 4.2.

Figure 4.2 explained

1. Subsystem (1) → Subsystem (1') *now outside TS but within Super-System,*
2. TS (2) before convolution → TS (2') after convolution,
3. Super-System (3) before convolution → Super-System (3') after convolution.

Some examples toward this:-

In future, say by 2025, in TS = car, only seat will remain inside car. All another parts would have joined road. These parts would 'manifest' as specialized nodal systems of energy supply, control, etc. Salient points: (1) super-system in this case is composed of car (TS) + roads + traffic signals and so on. (2) With car as TS, seat is the WU. It is the seat that acts on product, viz. man, and displaces him. (3) In many European technical universities, conclusive research on futuristic transport is directed toward equipping roads with engines. Roads in their models appear like a chess grid. Vehicles, akin to chess pieces are mere enclosed seats, just enough to contain one adult.

Modern society can be considered a super-system: system of collective usage (nodal, central), to which each TS constantly or periodically turns to. Elements of society, viz. humans, are technical systems (TS) here. Functions of super-system are: commutation, service, power supply, scanning, control, etc. This is a case of multiple TS under umbrella of a super-system.

Please refer Fig 4.2. Initially 1' fulfills same function as 1. However with time, number of TS increases—population growth. Quantitative changes inevitably lead to qualitative changes. With time differences between functions carried out by 1 and 1' grow. 1' gets more and more efficient.

This phenomenon works at higher level too. Consider several super-systems like energy supply, medical facilities, transportation, etc. under umbrella of a super–super-system. Application of type 1 convolution is left as an exercise for readers.

Finally, there is further gain in MUF (in terms of number of functions) that results from this type of convolution. As we know, the subsystem eliminated from TS joins as a specialized system into super-system. After joining, it integrates 'well' into super-system. A new system property or quality emerges due to this integration. This property or quality boosts MUF further. Some examples testifying this:-

The first telephone apparatus had switching units, sources of energy supply, and wires for connection with every other subscriber. In 1878, in New Haven, USA, the first telephone station appeared in world. It had type-setting switching field with plugs for connection between subscribers. Such local networks quickly spread in cities. Then the channels of connection between cities and countries appeared. Intermediary amplifiers, automatic switchboards and many other devices were needed. Modern phones are equipped with memory, answering machine, caller identification, fax, etc. Expansion process in terms of addition of new functions in this system continues in mobile phone connections. Simultaneously, process of convolution of subsystems occurs and displaces them into super-system. For example, the network of the artificial satellite absorbs a lot of technical systems in above-ground telephone networks; the necessity of nodes, switchers, amplifiers, cables, etc. in telephone stations falls off. Telephone network is of the many information systems available. Other such systems are radio, television, Internet, postal mail, etc. All these would join and emerge as united informational system, probably by end of this decade, if not earlier.

The second way of convolution—the development, mainly miniaturization of all subsystems contained in a given TS without displacement of subsystems into super-system. The peculiarities of this direction of idealization:

(a) M, D, and E decrease due to the miniaturization; sharp reduction of dimensions (D) decreases M and E as consequence.
(b) MUF increases due to the increase of the accuracy in functioning. Reasons: (1) length of ties decreases. This decreases the probability of mistakes, (2) required power decreases; 'E' of part of MDE falls. Also, many harmful effects disappear.
(c) The quantity of the elements of system stays unchanged up to last moment. But last moment consists in joining of subsystems into common functional mono-system.

The most characteristic example of mini and micro miniaturization in the engineering is development of electronics in twentieth and twenty-first (ongoing) century. The following illustration of this process is widely known: "If a Rolls-Royce in 50 years was improved by the same rate as computational machines, then this luxurious automobile would now cost two dollars, have engine with capacity in half cubic centimeters and would consume gasoline in 0.001 of cubic millimetre per one kilometre."

Sharp MDE in electronic systems has been in accordance with following chain: separate details—assemblage—micro assemblage—microelectronic circuit—highly-integrated chip—SBIS. The elements basically did not change all the way: it was the same set of resistive, capacitance, semiconductor, and inductive elements. Only in last several years, due to growing of electronic blocs in the form of the single crystals and assembling on the basis of biochips, signs of transition to basically new elements have appeared.

Consider future convolution of printing plant like this. Chosen book is printed in the presence of customer directly in bookshop. Text and illustrations are read from flash drive. In next few minutes, pages are printed out by a high speed laser printer and then bound by automatic binding line.

Suggested reading: Nanotechnology developed by Eric Dreksle of MIT, USA

Third way of convolution—convolution of technical system into single subsystem, mainly into subsystem containing working unit. Convolution of this kind can take place by four modes:

(a) Subsystem accepts carrying out of the function of some substance of TS. This substance is excluded from TS.
(b) Conjunction of two subsystems into one. One subsystem disappears.
(c) Conjunction of several subsystems into one.
(d) Convolution of TS into one of its subsystems.

Subsystems frequently have properties similar to properties of substances already been used in this TS in another of its part. It is necessary to displace this substance out 'having entrusted' the carrying out of its function to subsystem. If any subsystem does not have this necessary property, then it is mandatory to change it in the required direction. Several examples are given below:

In Australia, a solar photo-electronic sensor is made in the form of the tile from transparent plastic with plugged in photosensitive cells. Fastening them on roof is similarly to fastening of ceramic, cement, or steel tile. This tile functions as normal constructive tile element and additionally generates electricity.

In Japan, electric batteries with thickness no more than 0.1 mm are developed using firm electrolyte. It is suggested to place these batteries in the case (cover) of device or apparatus.

In Japan, home television antennae in the form of the wall calendar were released. Pictures were printed by metallic paints or were made using thin aluminum foil.

In Russia, kitchen-range without heating rings has been developed. The bottom of metal cooking utensil functions as heating rings. A thyristor converts frequency of AC from 50 Hz to 20 kHz. Electric current of raised frequency is delivered on primary transformer winding, and as secondary winding the bottom of cooking batteries is used. Efficiency of about 80 % is achieved. In contrast, traditional method, i.e. with coils, gives no more than 20 %.

At joining of subsystems someone of them becomes the 'main'. It accepts carrying out of additional functions from other subsystems. If one of subsystems is working unit, it definitely becomes 'main'. It always remains and its improvement continues. Others subsystems as though pulled together to WU; they remain close-by in its boundary layer 'flow' together with WU, so to say. Some examples of this:-

Control panel of the modern automobile is mounted on column of steerable wheel. Buttons are placed at such distance which lets us to get there by our fingers.

In Japan, lathe with electronic bloc of speed adjustment in which electric motor of constant current is combined with main spindle was created. Driving gear, tooth gears, shafts and muffs disappear.

Outboard engine with electric drive developed in Japan. Screw and electric motor are maximally drawn together and carried on consoles outboard.

Built-in screw propeller are made in Europe. Powerful screw of the ship must have big diameter and low amount of turns. At the same time, usually electric motors have big amount of turns and small rotor diameter. Therefore, huge reducing gears and shafts are used to couple motor with propeller. Both motor and propeller are subjected to heavy alternating load. To join engine with WU (screw) is offered in this innovation. Rotor of motor is a screw in which a boss is made of constant cobalt-samarium magnets. Stator made in the form of the ring, covers the ends of screw blades. Sharp decrease of MDE and other Harmful Effects (noise reduction) is reached. It is also easy to change the direction of the ship. A sample with screw diameter 2.5 m and power 750 kilowatt was created.

In measuring systems, WU is a sensor. Therefore, convolution of a measuring TS flows in the direction of joining all parts with sensor. For example, integral sensor is a silica crystal in which sensitive element and electronic circuit of the signal formation are joined. Such sensors have low MDE and increased MUF than their previous versions.

Even prison has been idealized (convoluted)! In USA, because of overcrowded prisons in many states, the house arrest of persons not convicted with grave crimes, is rather broad spread. And for their control at home, modern electronic facilities are used. The facilities of the control are of two types—active and passive, accordingly for more or less serious crime. Active facility, presents a constantly working transmitter which built-in untaken off bracelet put on the ankle of the feet of criminal. In his apartment, receivers are established which in random time are turned on—they transmit signals through telephone channels to computer placed at police station. Mass of transmitter is mere 70 grams. To extract it from bracelet special instrument is needed, and in attempt to cut or to take off bracelet, transmitter starts radiating special signals. The apparatus of passive control includes hand or leg bangle and the equipment of automatic enquiry.

Extended example: Electric bulb: powerful example of convolution of two heterogeneous subsystems. Stepped idealization not necessarily in chronological order is presented below:-

(a) An economical electric lamp is patented in USA. On the inside of bulb surface, thinnest possible layer of silver sandwiched between two layers of titanium dioxide is coated. This trilayer does not prevent visible light passing through but reflects infrared rays. Thus a transparent mirror has been created. Its shape is identical to that of bulb. Since such bulb is purposely ellipsoid-shaped, the transparent mirror is also identically curved. IR-rays emanating from filament are reflected inwards. They are focused back on the filament. Filament gets extra heat apart from electrical heating primarily provided. Only half electric power is consumed by bulb to give same light intensity. Here main subsystem is filament. Filament is also WU of TS (bulb). Thus main subsystem of TS is WU. Auxiliary subsystem is optical coating on inside of bulb surface. Product is light. Between inner layer of bulb surface and filament, there is nothing. Only light energy is present in this space. With each consecutive idealization, filament pulls coating closer to itself. Also, many functions of auxiliary subsystem are transferred to filament.

(b) Year 1925, Inventor—P.V. Bekhterev, USSR. 'Electric lamp with indoor reflector' is developed. Here is brief of reasoning offered by author of invention. In existing lamps, light is distributed irrationally, i.e. radiated luminous energy is used ineffectively. This defect is corrected in significant measure by correctly using phenomenon of reflection of light from inner surface of bulb. Enamelled iron reflects light weakly while mirrored coating causes blooming and is not very efficient. Indoor of the lamp is free from dust, flies, and oxidative gasses. The author presents to place reflectors of different types: parabolic, bulging, concave, decorative, and even Holophane inside bulb. About the last one: Holophane glasses of transparent kind with prismatic cross-section were bent and put one by one. Inside surface of bulb, that faces filament is smooth, whereas outer surface of bulb is ribbed. Complete internal reflectance was achieved inside bulb.

Interlude: Idea of combining reflector with lamp and applying definite geometrical shape to reflector kept on progressing further without pause.

(c) In USA, lamp with heat reflecting shield focusing radiation on tungsten filament is presented. However, at temperature of 2600–3000 K tungsten began to evaporate and settle on the bulb. And the bulb became blind.

(d) In FRG in 1975, for decreasing the temperature of spiral coil of tungsten, a material radiating visible light on heating, is placed inside bulb.

(e) USSR patent, for the same purpose, tungsten powder is covered by a core of aluminum nitride.

Interlude: In some of above inventions, inventors tried to use the harmful energy of IR-rays to attain additional heating of spiral. They were at once met with the limitation of the extreme temperature that tungsten could bear before evaporation. It was profitable, not to raise temperature so much and so drop down the quantity of primary energy (electrical) supplied for heating of spiral. To achieve maximum efficiency, percent of initial radiation in IR range is enlarged while percent of initial radiation in visible range is lowered. It is interesting to note that physics cooperated intrinsically; at lower power supply, bulb changes glow from 'yellowish-whitish' to 'reddish-orange'. Reason: at lower operating current, filament attains lower temperature. Not only does the total quantity of radiated energy weakens, its blackbody spectrum of emission shifts toward lower frequency (or higher wavelength). Thus, a 60 W, 220 V bulb operated at 40 W (by giving lower voltage) not only emits less light, but its color is more reddish-orange. Had it been operated at full 60 W, it would glow yellowish-whitish. Thus, by lowering temperature of tungsten, share of IR radiation emitted increases.

(f) France, 1973, inner surface coated by screen that reflects IR radiation back to filament but transfers visible one. In the beginning, bulb radiates 5–10 % of visible radiation and 90–95 % IR-radiation. IR radiates in large amounts reverts to filament raising its temperature. With higher temperature, filament shifts to 'whitish', viz. more visible.

(g) USA, bulb in the form of ellipsoid and the filament situated on one focus of ellipsoid is created. This design ensures higher in the main part of heat energy is spread.

(h) USA, 1978, spherical bulb having smaller area surface than ellipsoid and one focus instead of two is created. Body of glow (filament or otherwise) with small dimensions is placed at center. The heat returning to body of glow should be small because the center of sphere is a point. Such body of glow limits lamp power.

(i) USSR, 1983, Patent 1 023 451, Fig. 4.3 (top three images) electric lamp contains bulb of optic transparent material in the form of ellipsoid of revolution. Its inner coated surface reflects infrared while lets visible radiation pass. As expected, body of glow is inside bulb. It is a spiral filament. There is an additional component, reflector, having form similar to bulb form; its concave side faces body of glow. The indicated ellipsoid is formed by the rotation of the ellipse on approximately its smaller axis of symmetry. At rotation of eclipse, lots of focuses like f1 and f2 form a closed circle f. Spiral body of glow, spiral by nature, is shaped as toroid and is configured such that it covers this circle f completely. At turning on lamp, the body of glow radiates visible and invisible (heat) rays. Heat rays are shown by solid lines while reflected ones by one dotted. Visible rays which enter outside the limits of bulb, viz. room, are shown by double line. Reflected heat rays return to the glow body in another focus that becomes an additional heat source. The spread of rays is any axial section along the spiral length is same. Perfect geometrical overlap and optical coupling raises efficiency of this device.

(j) USSR, Patent 1 083 253, Fig. 4.3 (bottom image), electric lamp, in which with the purpose of the increasing luminous energy and reducing requirements of accuracy during production, the element having a strong coefficient of absorption of IR-radiation is fixed. Pressed boron silicide, carbide, or silicon nitride is material for this element. Melting temperature of element can be up to 3000 K. At heating, this bulb radiates additional light without the feeding of additional energy. Here, the second body of glow is dielectric. Direction of thought in mind of inventor: may be tungsten filament takes up role only as radiator of IR-energy. MUF of visible light radiation fulfilled by another element. Functions can be divided between two substances. This excludes problem of quick tungsten evaporation.

(k) USSR, 1984, Patent 1 100 658, Fig. 4.4, lamp in which body of glow is encircled by grid absorbing IR-radiation and radiating visible one. Ratio of the volume of grid to balloon is 1:4–1:3 and ratio of volume of body of glow to grid is 1:3–1:1. The dimensions of openings in grid are 35–80 μ. Grid is made of metal oxides and has in IR-range blackness of 0.4–1.0.

Lamp works in the following way. The body of glow heats up to the 2,600–3,000 K, radiates 4–9 % visible and more than 90 % IR-rays. Encircling gas, for example, xenon transmits heat to the grid (2). Grid is made of zirconium, thorium oxides, or hafnium with cerium which absorbs IR and radiates visible light. If openings in grid are < 35 μ UV increases and in dimensions more than 80 μ IR increases. Rough layer on inner surface of bulb is for the dispersion of light.

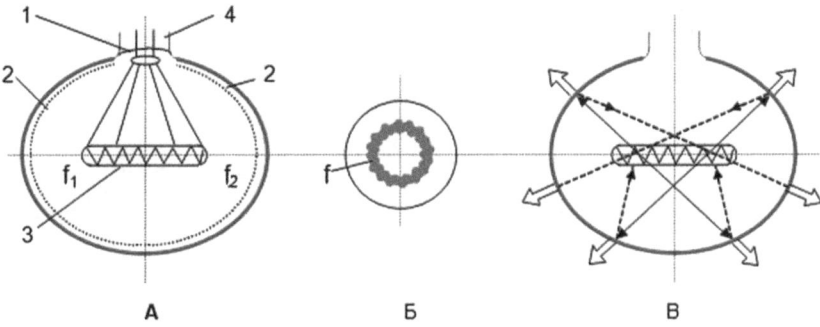

Electrical lamp. Patent 1 023 451.
1 – reflector, 2 – inner covering, 3 – body of glow, 4 – bulb.

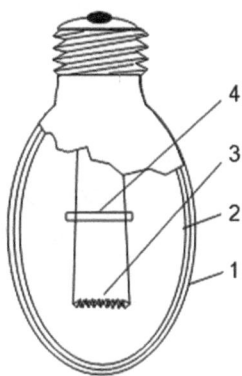

Electrical lamp. Patent 1 083 253.
1 – bulb, 2 – covering, 3 – body of glow,
4 – element for absorption in the IR
region.

Fig. 4.3 History of patents of electric lamps

(l) USSR, 1987, Patent 1 309 120, Fig. 4.5, lamp, in which inner reflecting bulb surface is formed by the rotation of two contrarily directed lengths of parabolas having general axis and general focus situated on the intersection of the axis of parabolas and the annular line of focuses. These two half ellipses are joined.

Fourth way of TS convolution—replacement of TS by ideal substance.

Convolution of TS into substance implies TS following either complete mode (1) or partial mode (2). Partial mode (2) occurs in three steps, (a), (b), and (c). The last step (c) of partial mode (2) is equivalent to complete mode (1).

Fig. 4.4 Continued history
of patents of electric lamps

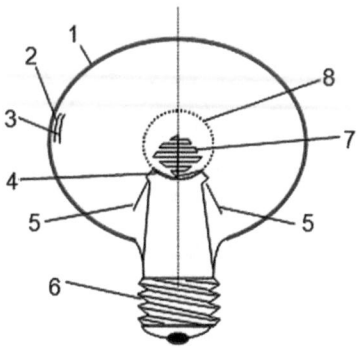

Electrical lamp. Patent 1 100 658.
1 - balloon, 2 – rough structure, 3,4 –
reflecting IR shields, 5 – element for
current delivering, 6 - socle,, 7 – body
of glow, 8 – oxide gird

Fig. 4.5 Advanced patents
in history of electric lamps

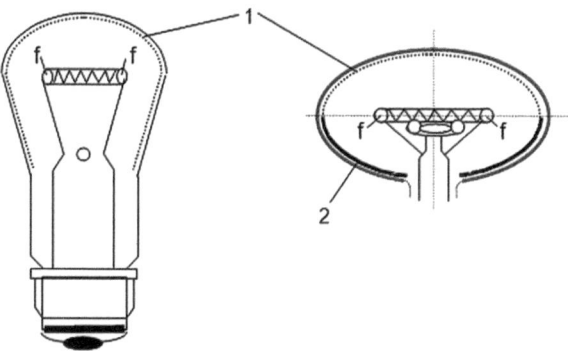

Electrical lamp. Patent 1 309 120.
1 – covering, reflecting IR radiation, 2 – mirror covering.

1. Complete mode: Vanishing of system but its function is fulfilled by a substance or,
2. Partial mode: substance being complicated, accepts carrying out of the greater and greater quantity of functions up to carrying out of function of entire TS. This can go in a several stages as below:-

   (a) One substance replaces functions of two or several substances.
   (b) One universal substance replaces several subsystems.
   (c) Replacement entire TS by ideal substance (IS).

Important properties of Ideal Substance that enable it to fulfill MUF of a high value are:-

(a) Self-organization;
(b) Independence of reply to the changes in exterior environment. Hence, exterior control action is absent or minimized.

Prognostic evolution of TS can be called directed idealization of substance made possible by means of the idealization operator. What is this idealization operator? Step by step, a subsystem or substances contained in a subsystem transfer their functions to just one substance. And finally into this one special substance, whole subsystem is convoluted.

Working unit being the most intensively developing part of system is stronger than others in attracting nearest substances and subsystems to it. For example, in system 'Rifle' working unit is a bullet. Nearest subsystem is barrel. One of the functions of barrel is rifling, viz. giving of rotation to bullet. We shall transmit this function of rifling to the working unit, the bullet. An example of possible solution: To use bullet-turbine with blades made of material embedded with memory of form. These blades are uncovered by heat of powder gasses. These blades twist the bullet. During flight, bullet cools down and blades are close again. Another function of barrel is heat removal. The example of solution: Bullet made of CPM is soaked in a substance with preset temperature of evaporation. Pores are blind, open only from back side.

Problem 9: A weathercock without mobile parts is capable to measure wind direction and speed. It consists of a silica chip with dimension about $0.5 \times 0.5 \times 0.1$ mm. Four thermocouples are placed on each side of the chip. Silica plate is heated uniformly from below. Silica plate contacts wind. From side where wind blows, plate cools slightly. Sensitive thermocouples detect this event. Higher the speed of wind on a surface, faster is cooling on that surface. Speeds ranging from 10 cm to 60 meters per second are measured with inaccuracy not exceeding 3 %. Electronics calculate the direction and speed of wind and carries out results on a small display. Immovable weathercocks can be a boon to yachtsman. Can this miniaturized TS be idealized further?

What is WU here? Screen showing direction and speed of wind. TS also has other parts like a source of energy supply, sensors (thermocouples), heater, the electronic scheme of signal processing, cable connecting screen with the unit of signal processing, etc.

The directions of further idealization:-

(a) Increasing of the quantity of useful functions like measuring temperature, humidity, time, etc.
(b) Exception of power supply and energy source subsystems. For example, the exterior source of feed should be replaced by source working on exterior environment—catalytic decomposition of air, moisture supplying oxygen and hydrogen as fuel element, solar radiation, the change of gravity characteristic at rolling, etc.
(c) All subsystems must be convoluted into screen, and then screen—into eye (or intermediary stage—into spectacles, into micro-projector for apple of the eye).

Problem 10: Safety from lightning fires in mining. Figure 4.6. Methane is the eternal companion in underground mining of coal. With air it forms a highly combustible mixture. Gasses pump out from mines and are thrown away into atmosphere through special high pipes (in engineering slang—'candles').

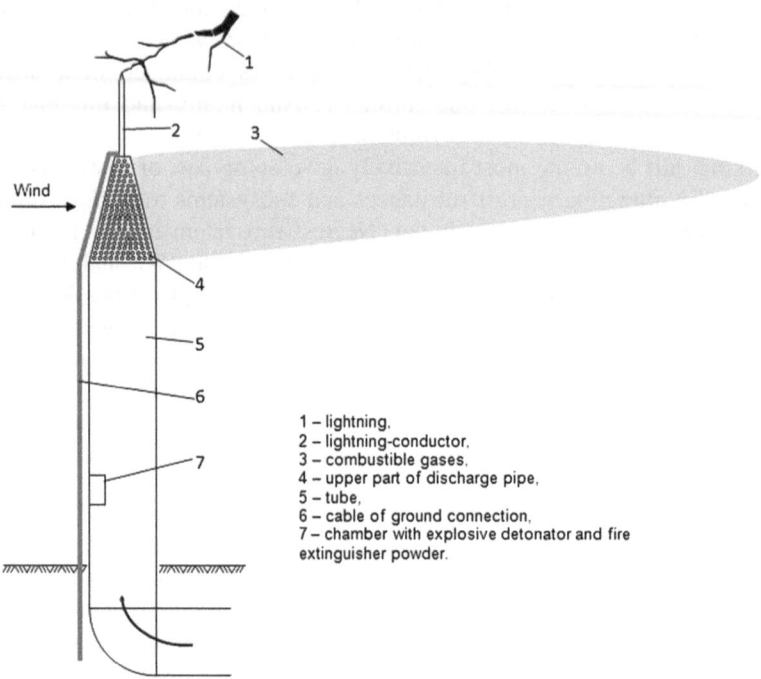

Fig. 4.6 Lightning powered lightning protection

These pipes should be high enough in order to prevent an accidental hit of spark from in moving gasses from some above-ground sources, like waste dumps burning frequently near the mines. But, the most dangerous is hit of lightning. Frequently, the lightning represents the branchy discharge, one of these branches can hit out-coming gasses. If unfortunate happens, flame with speed of sound directs on a pipe in mine. On an internal side of pipe, a chamber with explosive detonator and fire extinguisher powder is established to prevent this. For its operation, there is a system of automatic definition of the moment of hit of lightning. At moment of hit, a strong electric pulse is delivered to this detonator. The delay burns down and there is an explosion. For the control of working of this system, there is another control system. Both systems are powered from two independent energy sources, from electric main, and from diesel engine. How to raise ideality of system? That means to raise the reliability considerably and to reduce MDE.

A witty solution is proposed: it is necessary to use free of charge field present in super-system, lightning. Between ground cable and detonator, a connection strap is made. Lightning itself burns this package and quenches fire! Who says fire cannot fight fire?

Problem 11: Repair of electric mains, apparatus, and devices.

Electrician sometimes do not put on working clothes like rubber gloves, high boots, helmets, etc. and forget to check objects for presence of current before working on them. This is the main cause of electrocution-driven accidents. Strict instructions, verbal suggestion, posters, etc. do not always help. Pocket devices are used that radiate acoustic signal near the current source—higher the tension of electric field, stronger the signal. But this does not always saves; deterrent factors are strong manufacturing noise, loud conversation of workers, produced habit. It is necessary to propose an efficient method of protection from electrical shock: such that even if an electrician wants to but cannot a live wire. What is your suggestion?

Use of free of charge electric field in super-system is directed not on apparatus, but on worker. One of solutions: Bracelet on hand or electrodes in helmet. At entry into strong electric field, the electric current appears in metal under the action of electromagnetic induction and it acts on the skin of hands or heads. Unexpected prick of current in head makes worker sit down with fright. Second solution: Electrodes placed on the inner side of underclothes or in the area of shoulder or near elbow muscles. Involuntary reflex action itself pushes hands away near wire.

.

Chapter 5
General Scheme of TS Evolution in History of Technology

Human history contains limited openings and inventions, which shook the human population bases and gave powerful push to the development of civilization. Such revolutionary innovations are:

- Opening a fire,
- Invention of stone tools,
- Languish, written language, printing,
- Opening of electricity,
- Invention of information transfer methods without mass transition,
- Entry into space,
- Computer technology of information processing, and
- Biotechnology and genetic engineering.

We have to remember that all these events seem to us as the jumps of development only in historical prospect. Actually, these were the periods of more or less quick MUF increasing in any area of human activity. In these periods:-

(a) the big or small social forces and facilities were involved for the solution of the most important problems,
(b) a missing knowledge for solving the problems was created,
(c) openings and inventions made previously, which seemed superfluous previously and advance of their time, were considered seriously and used.

These advanced periods of TS development representing jumps or great increase of MUF were preceded by periods of TS stagnation represented by slowing down or total stop of MUF increase.

By the way, periods of such slow-down of development have been noted in many areas of engineering. We shall indicate only some of them:

S. Kwatra and Y. Salamatov, *Trimming, Miniaturization and Ideality via Convolution Technique of TRIZ*, SpringerBriefs in Applied Sciences and Technology, DOI: 10.1007/978-81-322-0737-5_5, © The Author(s) 2013

(a) A sewing-machine by E. Khou invented in the middle of nineteenth century performed 300 stitches per minute. The efficiency of modern machines is 3–3.5 thousands of stitches per minute if a cloth of natural fibers is sewed. For synthetic materials, such speed is too great; needle is heated because of friction, melting polymer.

(b) For 100 years, the average speed of metal cutting on lathe grew from 2.8 m/min up to 115 m/min; but from 1965, it has not risen practically.

(c) From 1965, the speed of the motion of conventional trains did not grow. The efficiency of rail transportation increased in fact that longer and heavier trains could be hauled at nearly same costs. We are not including TGVs or other magnetic levitated vehicles. But in any case, these sophisticated trains comprise no more than 5 % of total railroad transportation.

(d) Growth of factor of efficiency of electric generators stopped. It is approximately 30 % for heat-thermal electric generating plants and atomic/nuclear power-stations.

(e) According to the specialists, it is impossible and inexpedient to create generators with the power more than 2.5–3 million kilowatts.

(f) Electric mains approach to the limit of its possibilities; their tension (voltage) cannot exceed 2.2–2.5 thousands kilowatt.

(g) The limit of the increase of physical–chemical properties of traditional materials like cotton, wool, leathers, metals, alloys, armoured concrete, etc. has reached.

(h) Crop capacity of various, first of all, grain is close to extreme level.

(i) Active constraint in the parts of the intensification of technological processes is set by nature in animal breeding.

And so on. These and many other signs of slowing down indicate only about the approach of the period of accelerated development in each of these areas of human activity. Difficulties and contradictions will be overcome and new technology, new methods and technologies will provide MUF jump. Every MUF increasing is achieved by using knowledge received in the process of TS development. This knowledge is extracted from store-rooms of science or purposefully created under pressure of this problem. With the flow of time, MUF growth is slowed down and cycle is repeated.

Thus, there are grounds to speak about wave-shaped technological evolution. Assumption about some degree of symmetry within one wave is also possible: Out of 8 laws of Technical Systems' Evolution, some are more applicable in periods of expansion and while others are more applicable in periods of convolution Fig. 5.1. It is important to remember that different hierarchical levels of system (Substance-Subsystem-TS-Super system) can be seen in the various periods of development at one and the same time. For example, TS of a mobile phone might be expanding but simultaneously, super system of wireless telephony might be during convolution.

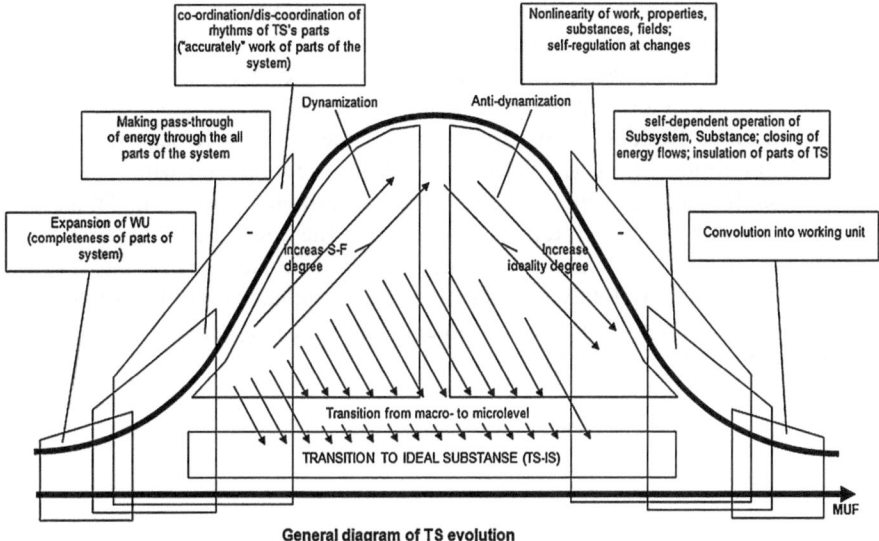

Fig. 5.1 General diagram of TS evolution

Beginning from the first industrial revolution and till today, the fluctuation in economic growth, in creative and innovative activity, in technological expansion is described by a few waves or cycles.

1. Inside the first cycle of development such inventions as steamer and loom are laid.
2. Inside the second—metallurgy, railway transport.
3. Third—chemistry, electricity, motor transport.
4. The technological base of the fourth cycle was: Electronics, high-molecular petro-chemistry, and aviation industry.
5. Ongoing fifth cycle is based on ceramics, semiconductors, laser technology, information and communication, space researches, biotechnology, artificial intelligence, nanotechnology, and space industry.

The wave rise begins from one or group of inventions which have appeared already in the period of previous wave of recession, depression. Innovation meets a strong counteraction. Furious resistance against implementation are explained by 'objective' reasons: there are no facilities, doubtful utility for today's necessities; frequently innovation is characterized by economic disadvantageousness; formed structure is broken, individual interests, and ambitions brake. However, more the resistance faced, more the potential energy for willing ascent accumulates.

Problem 12: In USA, patent 4 084 157, device of alarming fire reacts on the increase of temperature indoors and turns on an acoustic signal. This device consists of compressed spring held by easily meltable substance such as alloy of alloy of Nickel + Titanium, Paraffin, and etc. In case of fire, the substance melts, spring straightens and releases a valve on inflated balloon of air. Air moves into acoustic siren and an alarm signal generates.

This is not very reliable system because fire does not occur over years. During that time, spring can lose elasticity, substance can spoil (for example, get oxidized or decomposed). Besides, after turning on device, it is necessary to carefully prepare it for following cycle of operation. To set new balloon with compressed air, to press the spring, to fill it by easy melted substance and so on. Think what can be simplified in this system? How to raise the reliability of operation, to shorten the quantity of elements? In other words, it is necessary to improve this system or invent an absolutely new one. Here, you can have multitude of beautiful solutions because free of charge resource—heat field is appearing in fire. Let it self-signal! What is your solution?

Chapter 6
Convolution and Trimming via Convolution

What is Trimming? It is identical to convolution in results. Then what are differences?

(a) Convolution *occurs* in a technical system as a phenomenon; it is part of TS evolutionary wave. Trimming is a methodology purposefully applied to a TS to increase its ideality by achieving specific gain in MUF and/or to decrease one or all of M, D, or E from MDE by a desired amount.

(b) Trimming, if *done* correctly, is done via a system called 'Innovative Design' (ID). Several TRIZ tools including convolution methods are embedded in ID. Many a times, engineers cheat clients by following bogus methods: pruning is shown to client without latter realizing a latent HE has been attached to product or process. Thus, cheating in name of trimming, lean manufacturing, ecologically conscious carbon cutting is existing, if not rampant.

(c) Rules for trimming are reformulations of rules for convolution. Analogous to four types convolution, four basic rules of trimming are postulated.

(d) Trimming is generally done to a particular TS, a specific product or process of a company, in order to increase its commercial viability, increase its efficiency, etc. Results but more often steps of trimming are many times classified, not disclosed, retained under non-disclosure clause by just inventor and organization. Convolution happens (is obviously done by human inventor) to a generic TS. An example can settle this difference. In next few years, silent vacuum cleaners may appear with twice sucking power for same energy consumption, foldable into a briefcase, etc. That will happen via convolution; it will be recorded on TS–time graph. On other hand, manufacturer of a particular brand of vacuum cleaner may approach a design firm for reduction of volume by 20 %, noise by 30 % under some given operating conditions. The design firm will recruit a project management team which will trim this model of vacuum cleaner for client.

S. Kwatra and Y. Salamatov, *Trimming, Miniaturization and Ideality via Convolution Technique of TRIZ*, SpringerBriefs in Applied Sciences and Technology, DOI: 10.1007/978-81-322-0737-5_6, © The Author(s) 2013

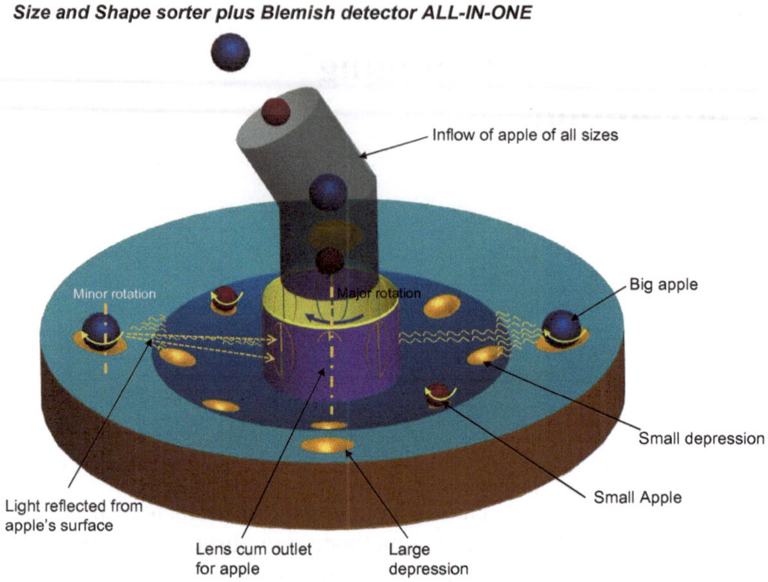

Size and Shape sorter plus Blemish detector ALL-IN-ONE

Fig. 6.1 Size and shape sorter plus Blemish detector ALL-IN-ONE

Case Study 1: Direct Invention of a Highly Convoluted TS

New *TS* is an apple multi-sorting machine

MUF is to separate apples on basis of sizes and blemishes

Previous TS are too expensive, heavy, cumbersome, and incapable of multi-task separation. So MUF is poor while HE is plentiful.

Working: Figure 6.1 for overall machine, Figs. 6.2 and 6.3 for two states of inner disk. Periphery of inner disk has uniformly spaced slits which can be closed or opened by sliding double doors. These doors are made of half lenses. In open position, both half lenses (half doors) disappear inside disk. When closed, both half lenses (half doors) snugly press other, forming a complete lens. In other words, open slit functions as normal open slit; mechanical articles, apples here, can cross it. On other hand, closed slit is a mechanical barrier but optically transparent. Thus, slit doors participate in two kinds of Su-fields: mechanical and optical.

Initially, slit-doors are closed. Apples of mixed sizes fall through a hose into inner disk. Either inner disk is alone set into rotation or both inner and outer disks are rotated together. In latter case, they may have identical or differential rpm.

When disk(s) reach final preset angular speed, slits are opened. Due to centrifugal forces, apples are thrown radially outwards. They cross the open slits to reach upper surface of outer disk. This surface is concaved slightly. On surface of outer disk, there are two (or more) sets of small partial hemispherical depressions. Depressions constituting outer set are larger; they belong to hemispherical profiles of larger diameter. In contrast, depressions belonging to inner set are part of

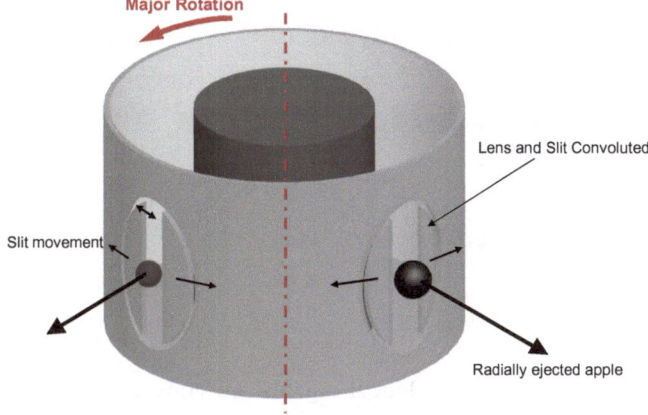

Fig. 6.2 Size sorting by preset slit gap

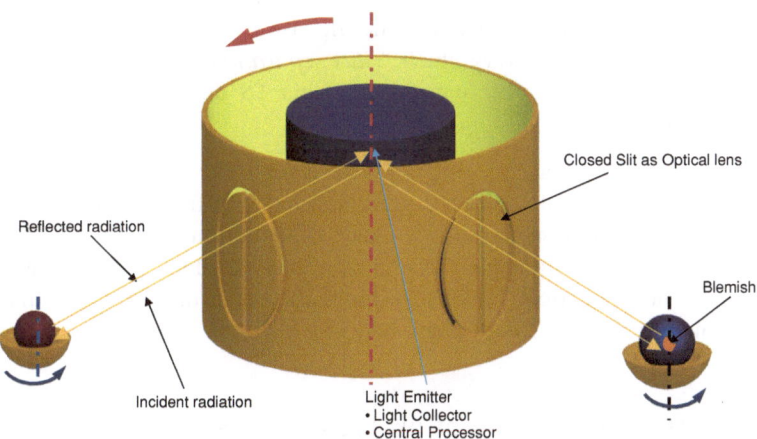

Fig. 6.3 Convoluted slit

smaller hemispheres. Apples that are small in size face smaller depressions. Apples of large size 'over-run' nearer small depressions to settle in further-placed larger depressions; their radius enables them to cross-over smaller depressions. *Separation of apples by size has occurred.*

Angular speeds of inner and outer disks are now equalized. Both can be brought to rest too, their speeds in this case being zero. Slit doors are now closed. They turn into convex lenses. Light beams are sent from a luminous source located at center of inner disk. These light beams pass through lenses, reach apples, and undergo reflection. Reflected rays once again pass through these same lenses to be collected

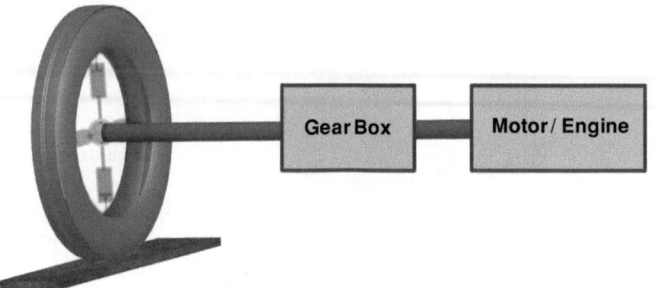

Fig. 6.4 Subsystems of a vehicle as TS

by solar cells or photodiodes. Optical signals are converted to electronic ones. These are then forwarded to a central processing computer.

The depressions, both small and large, are now also set into rotation. So apples, both large and small, start rotating along their own axes. This way, the lens can 'view' complete spherical surface of apple. In case, an apple has uniform color and little blemish, reflected light (hence electric signal) from that apple will have a nearly uniform graph with respect to time. In case of high blemishes, light output (hence electric signal) will vary much. *Thus separation by blemish has occurred.*

Following convolutions are identified in this TS development from conventional equipment.

(a) Subsystem of optical imaging and subsystem of slit closing-opening are replaced by one ideal substance. Ideal substance in this case is a split lens of optically transparent material like glass. Convolution of type 4 is in place.
(b) Subsystem of rotating apples around own axes develops. It is miniaturized in form of hemispherical bowls cut on main outer disk (rotating platform). Convolution of type 2 is in place here.

Case Study 2: Trimming of an Existing TS

TS is a vehicle, e.g. car. In this case, MUF would have contribution from parameters like cruising speed, maneuverability, safety, comfort of ride, etc. In MDE, M is mass of vehicle, D is dimensions of vehicle, while E is fuel consumption + human energy spent in driving (fatigue).

Trimming is done by applying convolution of 2nd type discussed in Chap. 5; one, more, or all subsystems of TS contract. Subsystem of gearing is eliminated and its function transferred to subsystem of wheel. Subsystem of suspension is also eliminated and its function transferred to sub-system of wheel. Overall, two subsystems are lost and only one subsystem is complicated at their expense (Figs. 6.4, 6.5).

Variable radius wheel replaces standard fixed radius one. In standard wheels, radius is almost fixed. Almost, because minor variations do take place due to fluctuations of air pressure in tire—but these are ignored. In our new wheel, radius

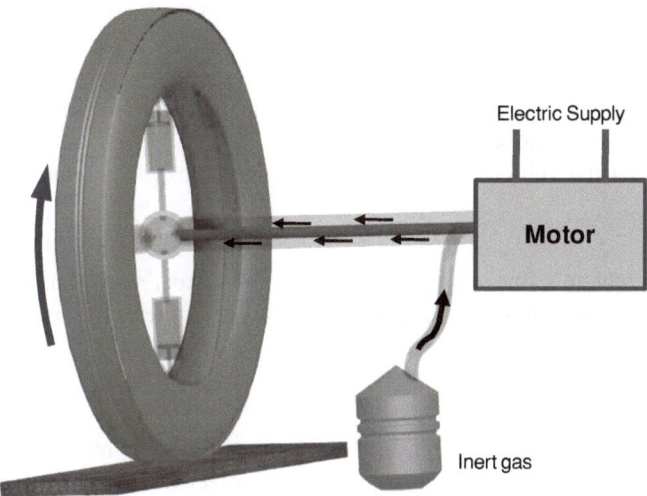

Fig. 6.5 Elimination of Gearbox

Fig. 6.6 Wheel radius set by gas pressure

can vary by great extent: ratio of maximum to minimum can be 3:1. The redesigned wheel consists of a small central hub, several spokes with every spoke broken to interlude a piston-cylinder mechanism, highly inflatable non-pneumatic tire outside. Other parts of wheel subsystem are: hollow axle with hollowness acting as air pipe, an air pump, controls, etc. Air pump delivers required air via hollow axle to piston-cylinders of all spokes. Pistons expand conforming with air pressure generated by air pump. Expansion of pistons lengthens spokes, which in turn expands outer rubber tire to desired radius R (Fig. 6.6).

Gearing subsystem is totally absent; its function has been fully transferred to wheel subsystem without error. Thus, drive subsystem is directly coupled to wheel system without any transmission (or gearing) subsystem. Drive subsystem shown here is electric motor based, though motor can be easily substituted by a standard gasoline engine too. Due to common axis of wheel and motor, both would have identical angular speeds at all times.

Use of linear and angular mechanics with several approximations is done below. Aim is to render results of convolution with fidelity rather than offer a dynamically precise numerical solution!

What about the vehicle's speed? Simple kinematics gives relation as:

$$V = \acute{\omega} \times R,$$

where, V is vehicle's linear speed, $\acute{\omega}$ is wheel (or motor)'s angular speed, and R is wheel radius at that time. We additionally assume a given or constant value of $\acute{\omega}$ initially.

$$\text{Thus, } V \propto R$$

Assuming, motor delivers constant mechanical power at all times, say P. Also suppose, this power P is 100 % transferred to wheel (losses in this transmission, even if there, would be negligible). Angular energy theorem applied to wheel gives:

$$P = \acute{\omega} \times \Gamma$$

where, Γ represents angular torque available for traction of vehicle.

Traction force transferred from wheel to ground, F need to be included now.

$$\Gamma = F \times R$$

Combining all linear relations stated above, we obtain:

$$F = (P / \acute{\omega} R)$$

With P, $\acute{\omega}$ as constants, $F \propto 1/R$

In words, Force and Radius emerge in inverse proportional function, while Speed and Radius emerge in direct proportional function (see first one of these equations).

When vehicle is started, air pressure is kept at minimum. Wheel has small radius R at that moment. With low value of R, a high value of F is generated. With low R, V is low too. This is just what we want. This is just what 1st gear achieves in a vehicle.

As vehicle speeds up (coming in mechanical equilibrium with wheel), air pressure is raised in pistons of spokes. Wheel expands to a higher radius R. With increased R, V increases but F decreases. This is what we want. This is just what 2nd gear achieves in a vehicle.

The next higher gears are 'engaged' in this way.

Harmful Effects (HE) emerge after this convolution.

1st HE: In normal vehicle, ground clearance H and wheel radius, R both are fixed by manufacturer. Many a times, manufacturer sets $H = R$ also. We assume this equality is respected. Figure 6.7 In our convoluted vehicle, vehicle radius, R, is

Fig. 6.7 R = D case

by definition a variable. But *D* should not change during motion of vehicle; you would not like to travel in a vehicle where your vertical elevation is speed dependent! Thus $R \neq D$ in this case; *R* may vary but not *D*. How to achieve this? One possible suggestion is illustrated in Figs. 6.8, 6.9. Wheel axis is held from roof via a piston-cylinder reciprocating mechanism. Part of air or inert gas fed to wheel is diverted to this dynamized suspension subsystem. Notice that this 'other way round' suspension subsystem can replace conventional suspension of wheel from chassis.

2nd HE: Radius of wheel has direct bearing on resistance to rolling motion. In case of variable radius wheel vehicle, dynamical relations have to be reformulated keeping *R* as variable (not a constant).

Extra benefits of this trimming:

1. Spokes of wheels are not rigid rods anymore—they act like springs with some stiffness. Wheel as a whole has a new property: diametrical elasticity along all possible diameters. Compliance of wheel to sudden, localized bumps on road improves.
2. Conventional suspension subsystem of vehicle may be eliminated or trimmed as part/whole of its function is transferred to wheel subsystem (point 1 above).

It is often observed that when TRIZ is applied to a TS to convolute it, desired results are supplemented by some extra, free and more desired results. A latent power of TRIZ is yet untapped. Before moving on, we identify convolution(s) and its type applied in trimming above. Type 2nd occurs with wheel subsystem developing to fullest. Note that miniaturization is a must for development. Here development can be labeled as sophistication without complexity.

The sub-subsystem of air pressure supply is eliminated. Its energy consuming component, air pump vanishes too along with. Subsystem of wheel takes on the function of sub-subsystem of air pressure supply. Although it takes on this burden of function, it gets simplified. Further convolution occurs. Newer wheel consists of two chambers with inside chamber made of rigid cylindrical substance while outer wall of outside chamber made of flexible cylindrical substance. Two chambers are

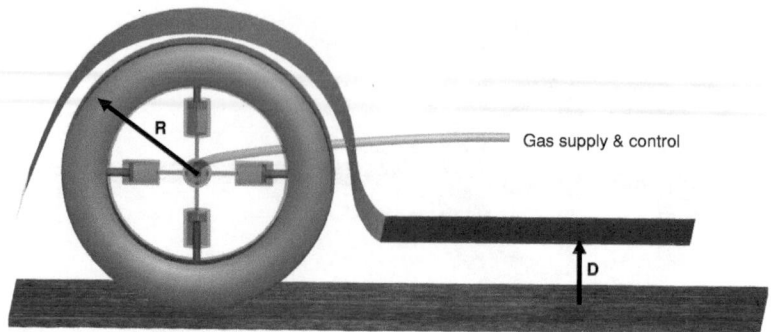

Wheel Radius ,R ≠ D , Vehicle Clearance

Fig. 6.8 R ≠ D case

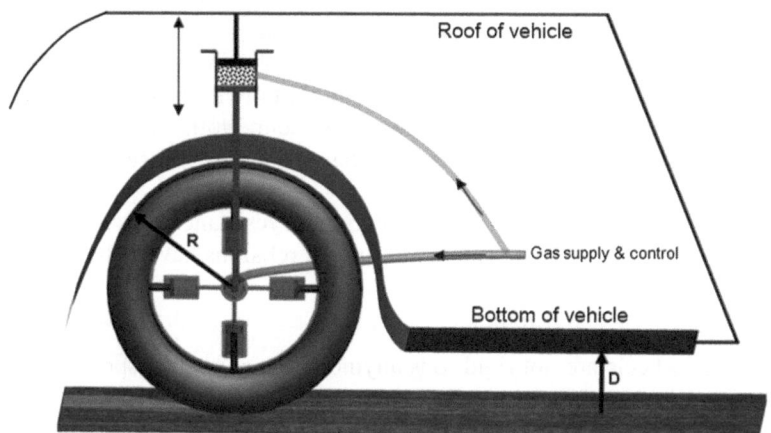

Wheel Radius ,R ≠ D , Vehicle Clearance

Fig. 6.9 R ≠ D case

Fig. 6.10 Super-trimmed vehicle

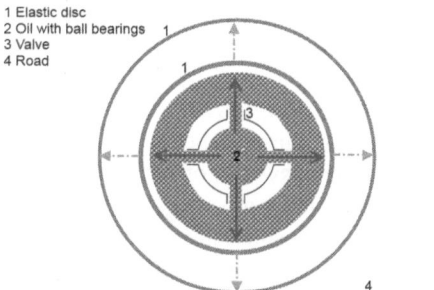

separated by valves. Inner chamber contains heavy inert oil with solid particles like ball bearings, etc. In next stage, we can choose ferroparticles of right size, mass, etc. as these solid particles; are we hinting at electromagnetic SFM?

At start of vehicle, oil with particles is totally contained in inside chamber. Wheel's outer chamber is deflated, like flat tire of car. As car picks up speed, centrifugal forces make oil and particles to move toward outer chamber through valves. Oil and particles together push outer wall of outer chamber in radial outward direction—wheel increases its radius R.

Angular energy of wheel has been used to gear it by itself. In this trimming, convolution of 4th type has occurred.

Trimming of Street Light Pole

TS is a Street Light Pole. Parts of TS:

(a) Strong pole made of metal or cement.
(b) Electric Bulb at top end.
(c) Electric supply socket at bottom.
(d) Electric wiring within or alongside pole's length.

Inventive Problem: Repair-ability and replace-ability of electric bulb tedious. Specific detail: Pole needs to be climbed up for detecting and rectifying even a minor flaw.

MUF of TS: To illuminate part of road

Brief version of Innovative Design* applied here. * Innovative Design given as Appendix.

As can be seen, Pole is source of only useful excessive actions (2) and Harmful action (1) Fig. 6.11. Thus, pole needs to be replaced/redesigned. Pole is composed of bunch of optical fibers which can be mutually twisted and bundled for added strength. Optical fibers transmit light with almost 100 % efficiency via principle of total internal refection. Twisting or other mechanical bending does not dilute this optical efficiency. Pole is lighter, but strong enough. Bulb is installed at bottom end of pole. Light travels from pole base to top end where it is emitted via suitable diffusers.

Convolution of pole by 4th type has occurred. Optical fibers in bundled and twisted formation take mechanical and electrical load simultaneously. MUF is maintained while MDE is reduced much. Ideality is obviously raised Fig. 6.12.

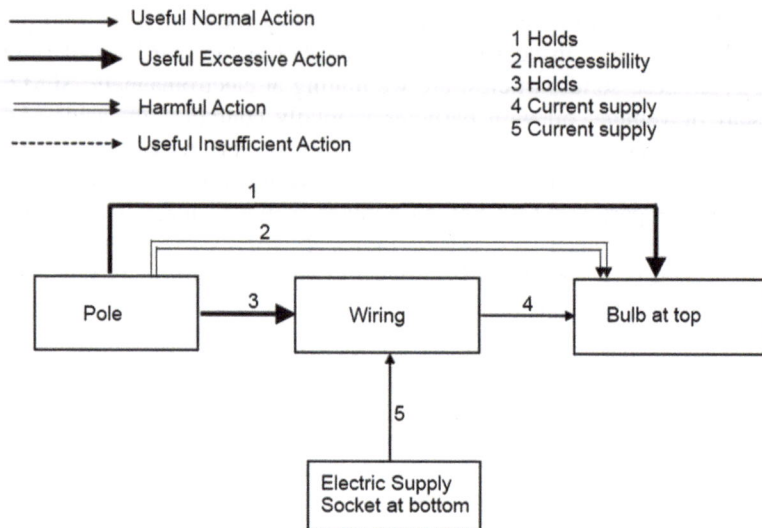

Fig. 6.11 Kinds of actions

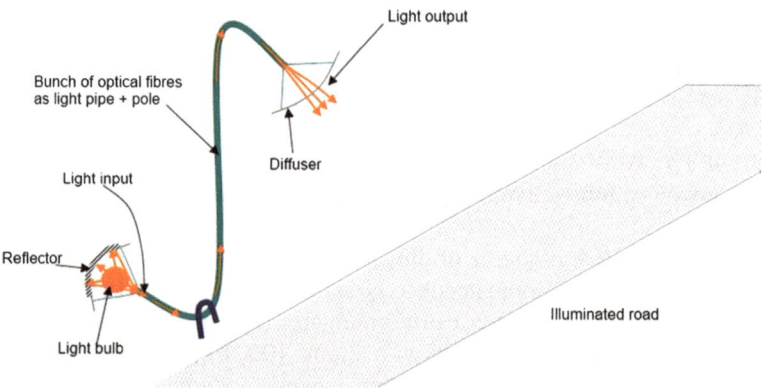

Fig. 6.12 Convolution of street pole light

Trimming of Lighted Screwdriver: Invention of Nano-LED Guided Screw-Driver

This invention relates generally to screw drivers, though it can be extended to other hand or hand held power tools. In particular, this invention is directed toward mini or micro screwdrivers.

Established Problem: It is well known that at times a person is obliged to place or remove a screw that is in a dark area, so that a separate flashlight must be employed to accomplish the task. Often there is not enough space for placement of

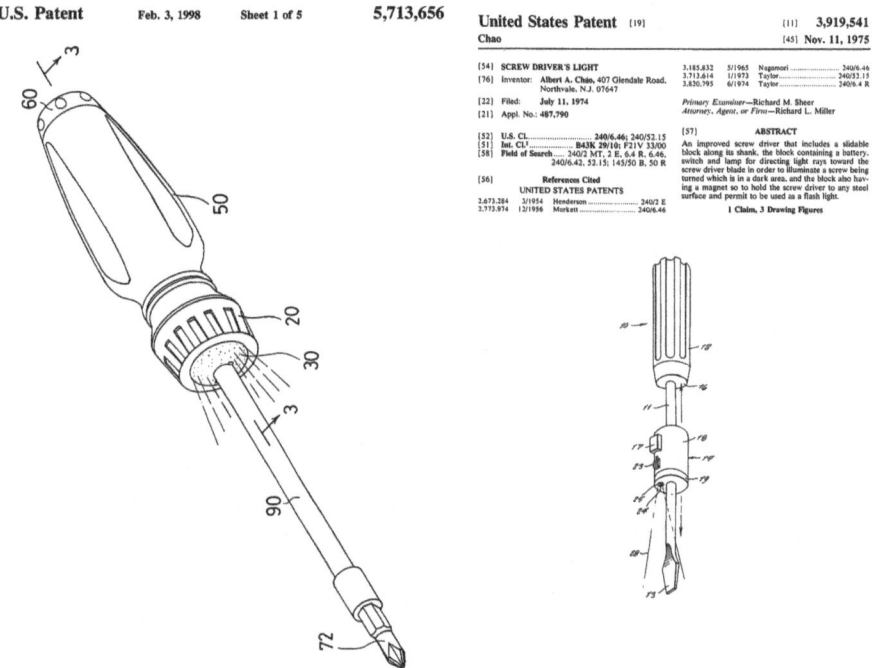

Fig. 6.13 Existing patents in lighted screwdrivers

a flashlight in the area, or the use of a flashlight necessitates a second person to hold it while a first person's hands are both busy holding the screw and the screw driver. This situation is accordingly in want of an improvement. Therefore, it is a principle object of the present invention to provide a screw driver that incorporates its own light for illuminating a screw being turned, and the vicinity thereof where the light is needed, and which thus eliminates the employment of external lighting means. What are shortcomings or limitations of existing patents on this?

1. Either additional space is taken by lighting subsystem.
2. TS is conventionally lighted screw driver. It is too complex.
3. Light source is way above screw so illumination is not bright enough on head of screw.
4. No adaptation for micro screwdrivers exists. Especially, miniaturization, viz. type 2nd convolution is prohibited or tough.

Figure 6.13 shows some existing patents.

We start with our trimming in a customized manner. Instead of pruning conventionally lighted screw driver, we start with standard screw driver as our TS and increase MUF by inclusion of illumination function. Strategy is to develop a convoluted TS with lighting function incorporated. This convoluted result can be 'considered' as trimmed version of cumbersome conventionally lighted screw driver.

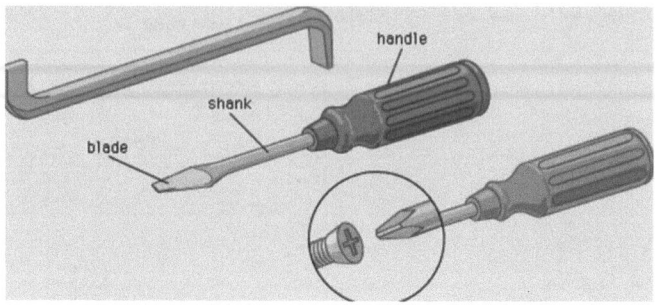

Fig. 6.14 Parts of a screwdriver

TS = basic screw driver, Fig. 6.14.

MUF = to tighten screws, bolts, similar machine parts.

Super-system = TS + bolt to be tightened (with machine part where located) + flashlight used at night (conditional adjacent technical system).

Environment = Sunlight or lack of it at night, gravity.

Inventive Problem = to add to existing MUF, another function of being able to work in poor light conditions without adjacent TS of flashlight.

MUF (desired) = to tighten screws in all-light conditions = MUF(1) tighten screws + MUF(2) provide light by TS.

1. If we desired, we could have left MUF of our TS unaltered and concentrated on Super-system.
2. In that case, Super-system could have fulfilled necessary MUF of light providing. For example, luminescent or radium coated bolt.
3. Also, in Super-system, MUF is tightening bolt in all light conditions. We could have used another physical principle of magnetism, electrostatic field, etc. that enables us to direct tip of screwdriver to exact face of bolt's head. We could have got away with light-emitting source altogether. Already magnetized tip of screw driver used to 'hold' it to bolt, screw, etc.
4. But we avoid Super-system route and confine ourselves to TS only.
5. In our TS = screw driver, add a subsystem SS(l).
6. MUF of SS(l) = provide light onto screw/bolt.
7. Parts of SS(l):

   (a) light source,
   (b) pathway for light,
   (c) battery,
   (d) electrical connectors,
   (e) switch.

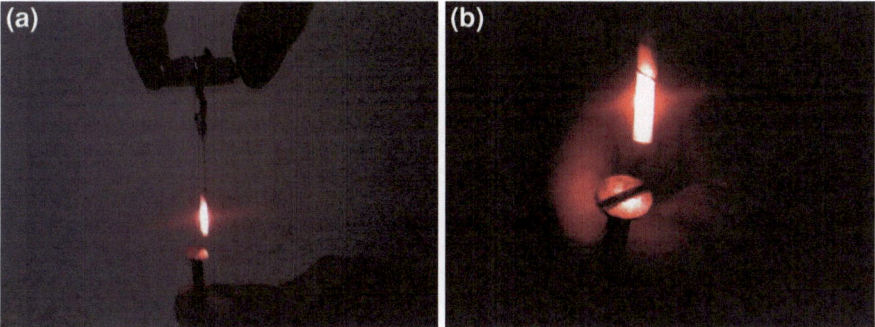

Fig. 6.15 Nano-LED employed screwdriver

Development of SS(l)

Choices made,

(a) light source = taken as nano-LED enclosed in shank, size: 0.8 mm
(b) pathway for light = shank of screw driver
(c) battery = displacing handle
(d) electrical connectors = enclosed and hidden in shank
(e) switch = can be placed anywhere

Modification of Parts of TS to Accommodate SS(l)

(a) Handle of SS displaced/replaced by Battery. Battery is rigid enough and can withstand stresses generated in handle of compressive and shearing kind due to pressing of screwdriver to screw and turning of hard nuts. Hence, it is feasible.
(b) Shank to be made of hollow tube that carries light efficiently by total internal refraction. Fiber glass or toughened glass possible. Strength of shank by replacing solid cylindrical shape to hollow cylindrical shape would not cause much disadvantage. Most chairs and tables have tubular legs of hollow beams. Advantage additionally is: weight saving. So MDE of TS goes down. Shank can enclose nano-LED and connection wires.
(c) Once hollow transparent tube replaces solid steel, LED can be moved up and down vertically inside shank by pulling or releasing wires. Application of Laws of dynamics growth of technical system.

Final technical solution photographed in Fig. 6.15. We leave it for readers to investigate type(s) of convolution occurring.

Fig. 6.16 Portable gauss
meter sold in market

Miniaturization of Existing Portable Gauss Meter

TS = portable Gauss meter (Fig. 6.16).

MUF = to measure electromagnetic field at a point in space. If it exceeds a limit, a warning signal is given. Note: MUF is only singular. Warning of exceed is a corollary of this MUF.

Shortcomings of existing Gauss meter:-

1. Needs external power source, like battery.
2. Is too big to be carried every time.
3. Basic aim is warning signal for common man. He is not interested in magnitude (strength) and direction of field. So display scale is useless for common man. For him, bare warning of field exceeding health standard is main aim.

TS' = Millimeter Gauss meter Fig. 6.17.

Super-system: Appliances like Microwave, Transformer producing harmful electromagnetic field, electromagnetic field in vicinity, human beings, Gauss meter, etc.

Subsystems are listed below:-

SS1. Subsystem of battery and power block totally wiped out. That is eliminated. External magnetic field itself becomes power source.

SS2. Subsystem of display (includes pointer, scale, back-end galvanometer coil, etc.) displaced by simpler subsystem consisting of single nano-LED.

SS3. Subsystem consisting of three different coils each on x, y, and z-axis displaced by single 3D coil.

SS4. Subsystem of warning (different colored lights, beeps, etc.) eliminated. Its function taken over SS2.

Functioning: EM field serves as detected signal + energy source.

Configuration: Three millimeter coils, one in each x, y, and z directions and nano-LED are all connected together in series or parallel.

Fig. 6.17 Gaussmeter-
highly miniaturized

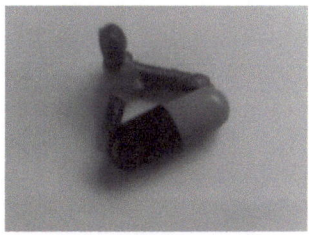

Fig. 6.18 Gaussmeter with
3-D coil

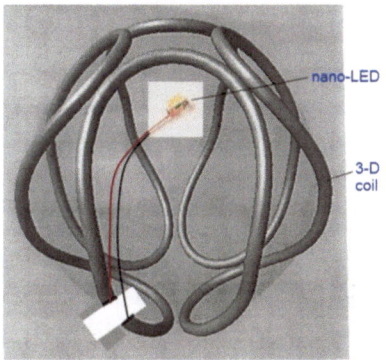

Figure 6.18 represents next possible convoluted state, say TS''. Three separate coils are displaced by a single 3D coil with its loose ends connected to nano-LED. Idle space within coil is partially used to make nano-LED sit. Flux is not much affected by displacing part of air by light source (material). Remaining space can be filled by iron fillings. Tightness will increase making system robust. Additionally, electromagnetic sensitivity increases because of higher magnetic permeability of iron.

Even further convolution is possible by replacing coil by a conductor carved in a semi-conductor. We call this TS'''. Readers are expected to think and work exact technical solutions for this. A step further can be TS'''' wherein detection occurs at human physiological level, say ions in blood stream signaling dangerous electromagnetic fields.

Here, trimming from TS to TS' was enabled by following types of convolutions:

(a) First type of convolution throws power supply subsystem from TS to Super-system. The external electromagnetic field now powers gauss meter, displacing batteries.
(b) Second type of convolution occurs in the development of field detection subsystem. Field detection subsystem is miniaturized by overwhelming proportions.
(c) Third type of convolution occurs in fact that subsystem of field detection, viz. three mutually right angled coils, becomes major, or daringly almost sole part of entire TS. It has almost replaced TS.

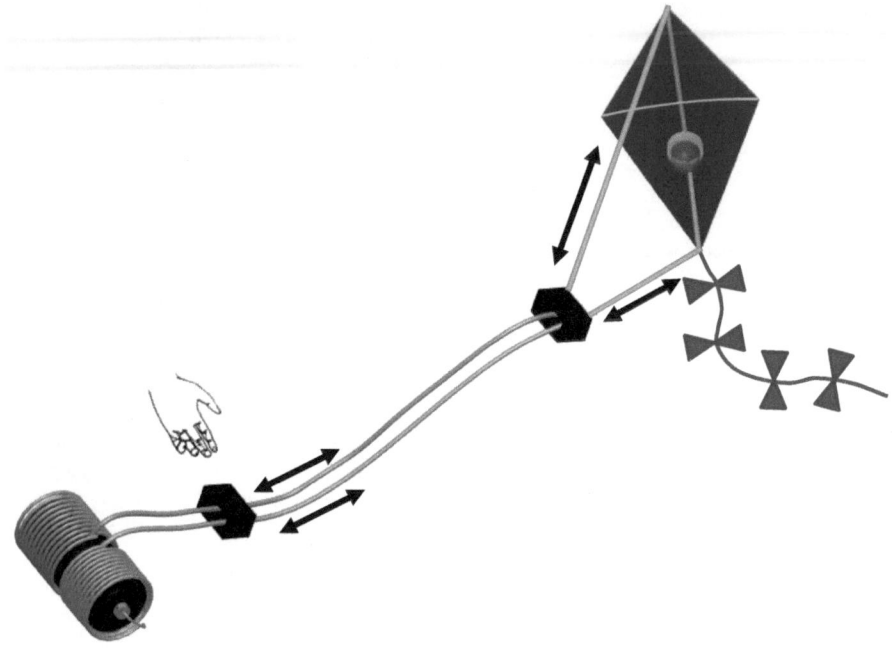

Fig. 6.19 Convoluted and dynamized kite-flying

TS' to TS'' is trimming by convolution of 2nd type; three mutually perpendicular coils are replaced by a 3-d coil singly. TS'' to TS''' would be trimming achieved by 4th and best type of convolution; an ideal substance does most things.

Lighted Kite Flying: Convolution

TS = Kite. Figure 6.19 is self-explanatory.

MUF = flying for entertainment, advertising possible too.

Addition to MUF = Light on kite, more dynamical control.

Subsystems modified: Spool by battery, single thread by double (twisted) electrical wires.

Subsystem added: Control to shift wires mechanically with respect to one another in order to dynamically control kite's position, angle of attack with wind, etc.

Readers are expected to identify convolution(s) with their types in this electromechanical kite.

Trimming of Washing Machine Using Innovative Design Methodology

TS = Washing Machine

Project Goals:

Main Goal: Only dirty portion of clothes are to be washed. Not subject entire cloth to full washing cycle because of one unwanted spill on a spot of fabric.

Other goals: Machine should be lighter, smoother in operation with some degree of automation, and feedback in-built.

MUF (required) of washing machine: to clean localized portion of clothes;

Definition of cleaning: to remove unwanted soil; to separate all foreign sticky/neighboring/embedded material from fabric. To get rid of psychological inertia, and to broaden our perspective, we move away from word 'washing'. Washing is just way of cleaning via hydro medium. So we call washing machine as cleansing machine.

To raise Ideality, we must maximize MUF and simultaneously decrease all Harmful Effects.

Restrictions: None

Acceptable degree of TS change: D: creation of new engineering

Type of inventive solution (1–5): Higher the better.

Measurement of MUF fulfillment:-

Optical images of pre-washed and post-washed clothes are compared. So physically, it is a luminous measurement. Optical subsystem may be added to TS for this.

Historic, Structural, and Parametric Analysis:-

Idea of washing clothes dates long back. For centuries, voyagers on sea washed clothes by placing dirty laundry in a strong cloth bag and tossing it overboard, letting ship drag the bag for hours. Principle was fine: forcing water through clothes to remove dirt. Washing machines in sense of a mechanism or sets of mechanisms appeared in 1,700s. From scrub boards to hand, powered washers to modern fully loaded automated washers-dryers, story continues.

Scientific Principle of conventional washing machine:-

Water (substance)—medium for chemical and mechanical transformations

Detergent (substance): Chemical additive

Physical action of scrubbing caused by agitation of water—mechanical field

Chemical action of detergent solution on soil—chemical field

Parts of a standard washing machine are shown in Fig. 6.20.

1. *Water inlet control valve*: Near the water inlet point of the washing, there is water inlet control valve. When you load the clothes in washing machine, this valve gets opened automatically and it closes automatically depending on the total quantity of the water required. The water control valve is actually the solenoid valve.

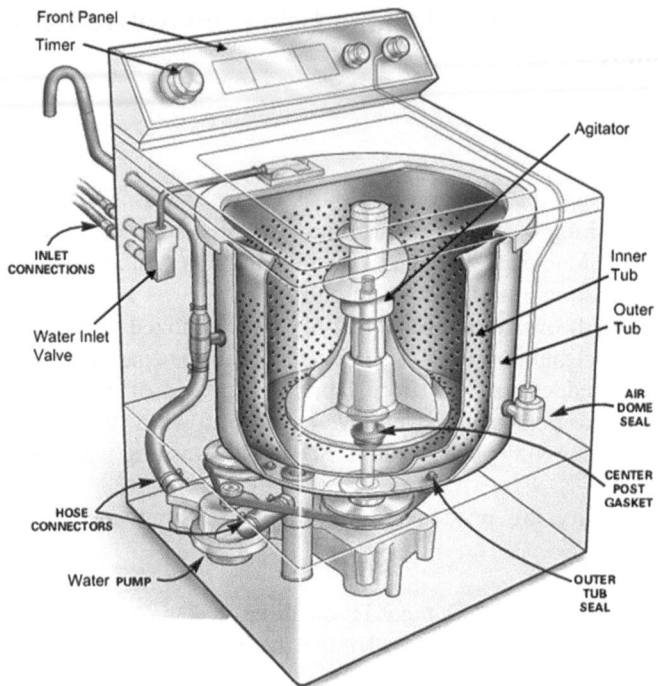

Fig. 6.20 Parts of a washing machine

2. *Water pump*: The water pump circulates water through the washing machine. It works in two directions, re-circulating the water during wash cycle and draining the water during the spin cycle.
3. *Tub*: There are two types of tubs in the washing machine: inner and outer. The clothes are loaded in the inner tub, where the clothes are washed, rinsed, and dried. The inner tub has small holes for draining the water. The external tub covers the inner tub and supports it during various cycles of clothes washing.
4. *Agitator or rotating disc*: The agitator is located inside the tub of the washing machine. It is the important part of the washing machine that actually performs the cleaning operation of the clothes. During the wash cycle, the agitator rotates continuously and produces strong rotating currents within the water due to which the clothes also rotate inside the tub. The rotation of the clothes within water containing the detergent enables the removal of the dirt particles from the fabric of the clothes. Thus, the agitator produces most important function of rubbing the clothes with each other as well as with water.

   In some washing machines, instead of the long agitator, there is a disk that contains blades on its upper side. The rotation of the disk and the blades produce strong currents within the water and the rubbing of clothes that helps in removing the dirt from clothes.

5. *Motor of the washing machine*: The motor is coupled to the agitator or the disc and produces it rotator motion. These are multispeed motors, whose speed can be changed as per the requirement. In the fully automatic washing machine, the speed of the motor i.e. the agitator changes automatically as per the load on the washing machine.
6. *Timer*: The timer helps setting the wash time for the clothes manually. In the automatic mode, the time is set automatically depending upon the number of clothes inside the washing machine.
7. *Printed circuit board (PCB)*: The PCB comprises of the various electronic components and circuits, which are programmed to perform in unique ways depending on the load conditions (the condition and the amount of clothes loaded in the washing machine). They are sort of artificial intelligence devices that sense the various external conditions and take the decisions accordingly. These are also called as fuzzy logic systems. Thus, the PCB will calculate the total weight of the clothes, and find out the quantity of water and detergent required, and the total time required for washing the clothes. Then they will decide the time required for washing and rinsing.
8. *Drain pipe*: The drain pipe enables removing the dirty water from the washing that has been used for the washing purpose.

Parallel areas in bio-engineering where cleansing is achieved

1. Dentistry. Daily brushing of teeth by humans, regular cleaning of teeth of animals without brushing. Power of saliva. In teeth cleaning, teeth are fixed and only solution (saliva + water + toothpaste) moves.
2. Fabrics of clothes. Like cotton. How is cotton cleaned prior to being woven into fabrics? So we analyze cleansing of same substance at earlier part of manufacturing process.
3. Other soap-based cleaning operations. Like floor, dishes, bathing, etc. Idea of natural cleansing of atmosphere by rains.

Technical Function to clean clothes $- \to$ Generalized technical function: to separate extraneous harmful substances like color, oil, grease, dust, etc from given substance (cloth) with aid of additives like detergent, softener, whitener (substances), and applied fields like mechanical (rotary), thermal, ultrasonic.

Problem 1: Fields applied are not accurately measured. They are not controlled.

Solution 1: Either use more controlled Computed mechanical fields or change from mechanical to electromagnetic for more control:

Solution 1.1: Floor cleaning by automatic robotic sweeping machines. They cover entire area without overlap. High efficiency $\to$ Let motion of water and/or clothes be controlled in real space–time using microprocessor- > Modern washers are microprocessor controlled but vaguely and at macro-level. More control possible at micro-level.

Solution 1.2: Electrostatic dust separator in pollution control. Installed in chimneys- > Let clothes and dust be oppositely charged and separated by ionic repulsion- > Woolens produce spark while wearing or unwearing. Further research required to use this effect in cleaning.

Problem 2: Substances used—water, detergents are NOT good differentiators.

Solution 2: Chemically superior substances needed. Either better substances or a better combination.

Solution 2.1: In dry cleaning, petroleum products are used instead of water- > Cleaning with alcohol or petroleum bases can be examined- > Washing machines with solvents as air, petrol can be constructed.

Solution 2.2: Carpet cleaning uses processes in sequence with different solvent or chemicals- > Can sequential washing machines be made? Even now, softener is added in latter part of washing cycle- > Washing machines with complex cycles with different chemicals in different cycles can be explored.

First some humor in sci-fi story! In his book "Travels with Charlie," the author (and Nobel Laureate) John Steinbeck describes a washing machine he put together in the back of his truck as he was traveling. If my memory serves me, it was a garbage with some water and soap in it, suspended between bungee, (that is, rubberized elastic) cords. As the truck rattled down the road, the vehicles shaking was translated into agitation of the garbage can. This proved to be very effective at cleaning the clothes within. Needless to say, it is also a pretty cheap solution.

Washing technologies that evolved over years. Most of the times, washing machines agitate dirty clothes in water with detergents. Mode of agitation was amplified (whether better or not, I am not certain) in more or less following order.

Vibrating systems – → Continuous Rotating systems- → Discontinuity (according to some cycle) with heat and pulsation added to rotation system → Dryer clubbed in: spin and heat in succession → Automation without feedback- > Automation with sensors and feedback Addition of 'experimental' enhancements like ultrasonic waves, high pressure water jets, cavitation bubbling, brushing, or batting

HEs

1. Impossible to wash clothes in heterogeneous portions like underwear, towels, dirty linen separately from main clothes like shirts, trousers, etc. Life of entire fabric unnecessarily shortened due to single dirty spot
2. Bacterial cleansing achieved in sun-drying absent (in dryer portion of washing machine)

Figures 6.21, 6.22, 6.23, 6.24, 6.25 are illustrations of Trimming achieved by Innovative Design Technology.

Washing substance M1
water + detergent + softener
Washing substance m1
water + dirt + residues
Dirty clothes m2
Clean clothes M2
M1 — → m1
m2 — → M2
We choose level 5 type of innovation
Working principle of washing may be revolutionized

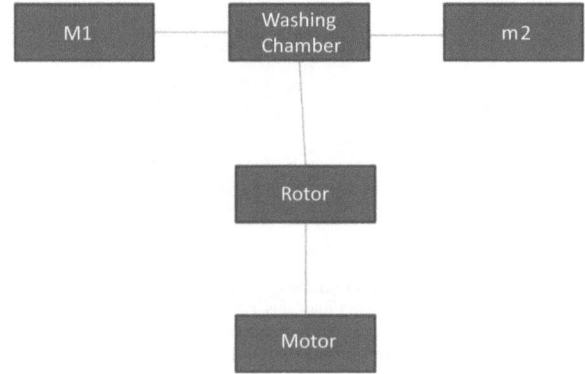

Fig. 6.21 Connections and Ties in washing machine

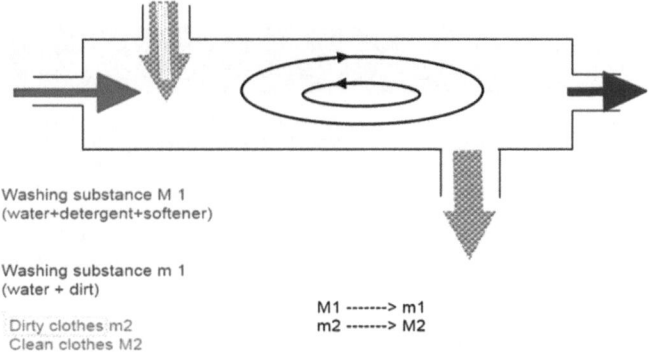

Fig. 6.22 Functional analysis of washing machine

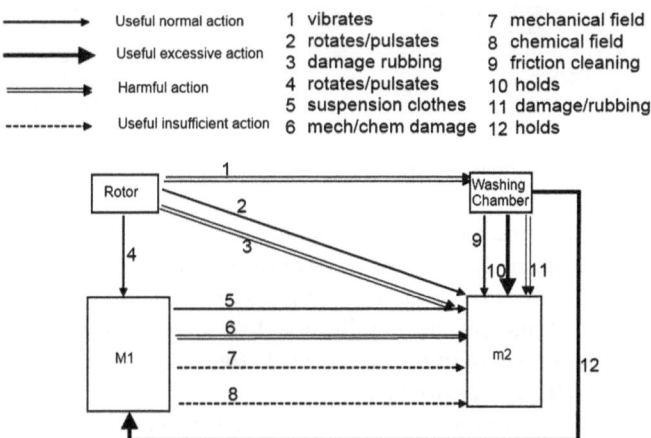

Fig. 6.23 Actions in washing machine

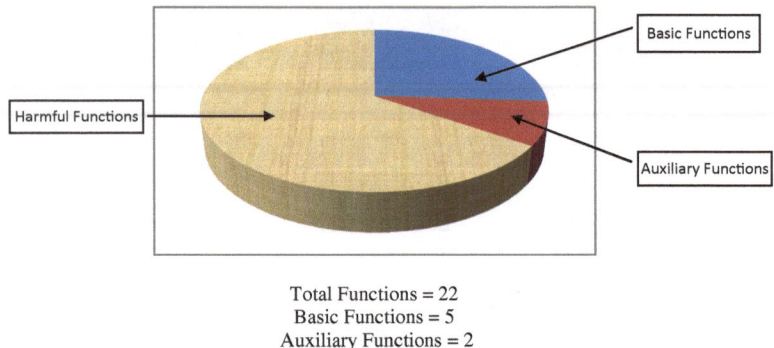

Total Functions = 22
Basic Functions = 5
Auxiliary Functions = 2
Harmful Functions = 15*

Fig. 6.24 Functions and their percentages

Element of TS	Functional Significance	Problem Significance	Cost Significance
Water + additives M1	Important for MUF fulfilling	Associated with 4 harmful effects overall	Medium comments: water is free resource but heavy & voluminous & detergent medium cost but to rotate M1 with varying angular speeds requires lot of power
Rotor	Not at all important for MUF fulfilling	Associated with 6 harmful effects overall	High comments: this part, though small-sized is run by expensive electric motor
Washing Chamber	Slightly important for MUF fulfilling	Associated with 5 harmful effects overall	Very High comments: this part is massive, voluminous, expensive to manufacture, install and service

Fig. 6.25 Cumulative functional analysis

Therefore, TRIZ principles may be applied to System

We do not attempt to change Super-system—it is too difficult and has boundaries with substances used in clothing, climatology, human biology, and culture.

Super-system: Atmospheric conditions (like heat, pollution) and human biology (like sweat producing glands) work together to make clothes dirty. They need to be cleaned regularly.

System: Washing machine, water, clothes, detergent, bleacher, softener and whitener.

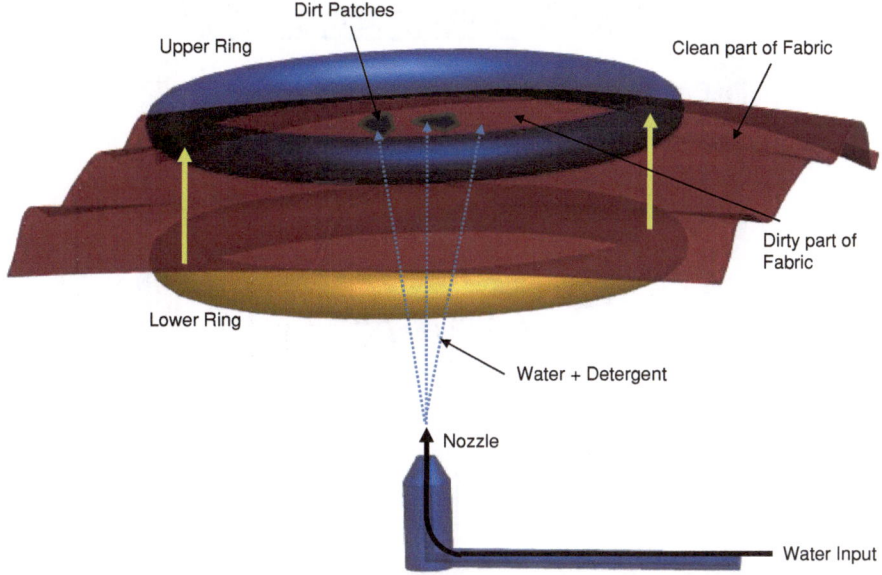

Dirt Patches

Upper Ring

Clean part of Fabric

Dirty part of Fabric

Lower Ring

Water + Detergent

Nozzle

Water Input

Fig. 6.26 One concept of trimmed washing machine

Subsystems of Washing machine as TS.

Inlet system for water.

Drainage system for disposing water + residues.

Detergent additive mechanism with timing control.

Additional devices like ultrasonic generator, heater, brush, etc.

Electronic control circuit for cycle setting.

Rotation/Pulsation mechanism—For Level 1 type, this needs improvement.

High rpm spin system for spin-drying.

Dryer with heat.

It appears that washing chamber and rotor be eliminated or trimmed to increase MUF to improvize washing machine.

Final Technical Solution: Figures 6.26, 6.27: *Semi-automatic water curtain washing machine with localized intensified cleansing.*

The double-decked machine's MUF is intensified cleansing of localized dirtiest or heavily soiled parts of fabric. This function is transferred by additional nozzles in 'water curtain' washing machine. These additional nozzles direct highly concentrated detergent solution at dirty spots. These spots are optically detected—color and luminous properties of spots are signal senders. The water nozzles 'follow' the optical sensors. This feedback leads to semi or full automation. Great decrease of MDE, increase of MUF, and consequent high degree of ideality value are final results.

What kind of convolution(s) took place in washing machine development? Dominant type is 3rd. Water-clothes intermingling subsystem becomes lean with rotor thrown it. This subsystem also eliminates washing machine body

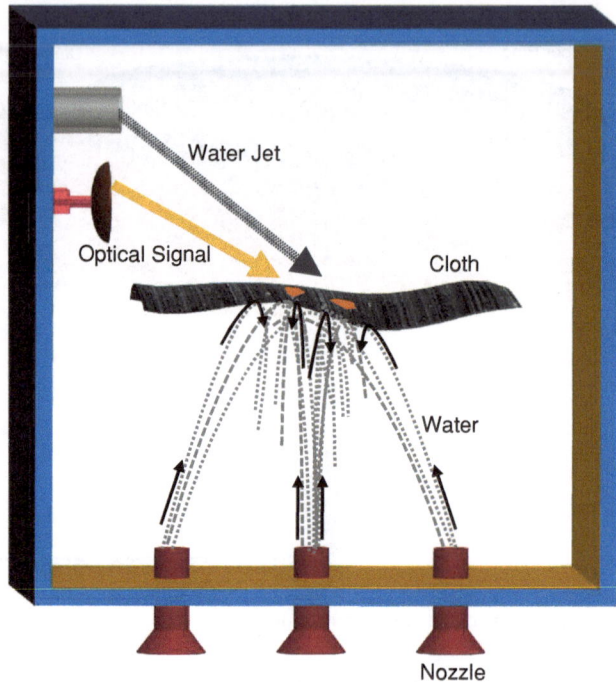

Fig. 6.27 Another concept of trimmed washing machine

subsystem—water curtain from jets isolates washing area from environment and this acts as physical body. Note that physical body is of water, a liquid here. Much lighter than metal for sure. Thus, water-clothes intermingling system almost functions like full-fledged TS.

Appendix
Teaching Convolution in Classrooms

Case Study of Developing a Non-spill Teapot

If a teapot purchased from market and one bought from an antique shop are given to us, I doubt we can differentiate which is which unless manufacturing date stamped on bottom is viewed or owner furnishes details. Especially, if a contemporary teapot is an esthetic, classy one. Has teapot then not evolved? Figure A.1.

Without entering into history of teapot design over ages, let us try to add a new function to its MUF. Existing MUF of teapot consists in holding hot beverage safely and pouring it into cups sequentially. Many shortcomings exist in modern teapots; each of them if challenged can result in a desire to increase MUF. We herein identify just one major shortcoming of teapot. Pouring without spill. Pouring when pot is full is an art turning into science when few cups are filled. A teapot filled to its brim is difficult to transport and when first cup is filled caution is exercised to prevent table cloth from getting stained. It would be good if mechanism existed that could control speed of teapot rotation; human hand by itself cannot articulate such sensitive angular velocity and displacement controls. This is similar to fine-tuning in old radios. While major tuning made needle close to a broadcast station and get voice, fine-tuning was a gearing that allowed needle to move ultraslow and get needle to the exact frequency.

Physics of flow from sprout of teapot is complicated. Without debating on it, we assume that it would be better if teapot rotated slowly at first when tea was full and rotated fast later on when tea was say half-full.

Designers would think of adding a new subsystem of gearing—they could do this by attaching a gear box between hand and teapot. This gearbox was then some kind of transmission between engine (hand) and teapot (working unit). But then gearbox has to be controlled. It must be in lower gear, say 1st gear when started and be shifted up to 2nd gear a little later. So one subsystem necessitates another subsystem.

S. Kwatra and Y. Salamatov, *Trimming, Miniaturization and Ideality via Convolution Technique of TRIZ*, SpringerBriefs in Applied Sciences and Technology, DOI: 10.1007/978-81-322-0737-5, © The Author(s) 2013

Fig. A.1

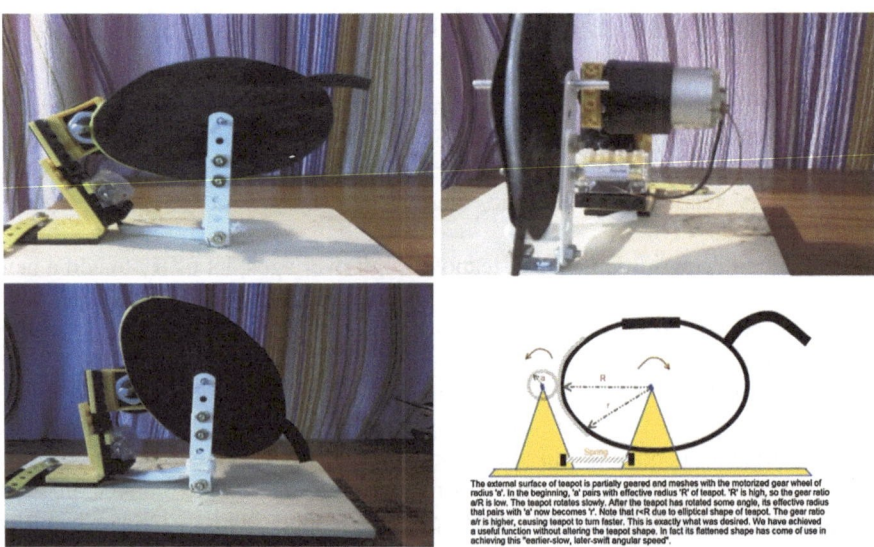

Fig. A.2

What is escape? Convolution. Teapot normally has an ellipsoidal shape. Its shape can itself render gearing. A constant speed motor meshes with outer surface of teapot, ellipse in 2-dimensions. At start, the larger radius of ellipse makes teapot rotate slowly. As ellipse rotates, its effective radius decreases, eventually reaching its minor radius. Teapot, thus accelerates, as its radius changes from larger to smaller one smoothly. The exact desired function of teapot angular velocity versus time can be set in terms of magnitudes and ratios of major to minor radii. In fact, other shapes can be experimented with both mathematically and empirically until mechanism design presents desired system characteristics during pouring. Figure A.2